AF581713

3D Printing for Tissue Engineering and Regenerative Medicine

3D Printing for Tissue Engineering and Regenerative Medicine

Special Issue Editors

Murat Guvendiren

Vahid Serpooshan

MDPI • Basel • Beijing • Wuhan • Barcelona • Belgrade

Special Issue Editors
Murat Guvendiren
Department of Chemical
and Materials Engineering,
New Jersey Institute of Technology
USA

Vahid Serpooshan
Georgia Institute of Technology
& Emory University
School of Medicine
USA

Editorial Office
MDPI
St. Alban-Anlage 66
4052 Basel, Switzerland

This is a reprint of articles from the Special Issue published online in the open access journal *Micromachines* (ISSN 2072-666X) from 2019 to 2020 (available at: https://www.mdpi.com/journal/micromachines/special_issues/3d_printing_tissue).

For citation purposes, cite each article independently as indicated on the article page online and as indicated below:

LastName, A.A.; LastName, B.B.; LastName, C.C. Article Title. *Journal Name* **Year**, *Article Number*, Page Range.

ISBN 978-3-03936-112-0 (Hbk)
ISBN 978-3-03936-113-7 (PDF)

Cover image courtesy of Murat Guvendiren.

Contents

About the Special Issue Editors

Murat Guvendiren is Assistant Professor at the New Jersey Institute of Technology (NJIT). He earned his B.S. and M.S. degrees in Metallurgical and Materials Engineering from the Middle East Technical University in Ankara (in Turkey). He was awarded his Ph.D. from Materials Science and Engineering at Northwestern University, which focused on adhesion in hydrogels and glassy polymers. His postdoctoral training was conducted at the University of Pennsylvania, focusing on stem cell interactions with dynamic and patterned materials. He was Research Assistant Professor at the New Jersey Center for Biomaterials at Rutgers University between 2013 and 2016. He joined NJIT in 2016, with a primary appointment in the Chemical and Materials Engineering Department and a joint appointment in the Biomedical Engineering Department. His research group focuses on the development of novel biomaterials (and bioinks) with user-defined and dynamic properties, investigation of cell–biomaterial interactions, and additive manufacturing (including bioprinting) of tissue engineering constructs and in vitro disease models.

Vahid Serpooshan received his BS.c. and MS.c. in Materials Science and Engineering at Sharif University (Iran, 2003) and Ph.D. in tissue engineering at McGill University (Canada, 2011). His Ph.D. research focused on the design of scaffolding biomaterials for bone tissue engineering applications. Following his Ph.D., he worked for 7 years at Stanford University School of Medicine as Postdoctoral Fellow (Pediatric Cardiology) and Instructor (Stanford Cardiovascular Institute). At Stanford, Dr. Serpooshan's research mainly focused on developing a new generation of cardiac patch device to repair heart tissue following heart attacks. The patch was successfully tested in animal models and is now in clinical trials. He also worked on enabling technologies for human–machine hybrid cardiac tissue using 3D bioprinting. In 2018, Dr. Serpooshan joined Emory University and Georgia Tech as Assistant Professor of Biomedical Engineering and Pediatrics, where his multidisciplinary team is now working on a variety of 3D bioprinting-based tissue engineering projects.

Editorial

Editorial for the Special Issue on 3D Printing for Tissue Engineering and Regenerative Medicine

Vahid Serpooshan [1,2,3,*] and Murat Guvendiren [4,5,*]

1 Department of Biomedical Engineering, Emory University School of Medicine and Georgia Institute of Technology, Atlanta, GA 30322, USA
2 Department of Pediatrics, Emory University School of Medicine, Atlanta, GA 30322, USA
3 Children's Healthcare of Atlanta, Atlanta, GA 30322, USA
4 Otto H. York Chemical and Materials Engineering, New Jersey Institute of Technology, Newark, NJ 07102, USA
5 Department of Biomedical Engineering, New Jersey Institute of Technology, Newark, NJ 07102, USA
* Correspondence: Vahid.serpooshan@bme.gatech.edu (V.S.); muratg@njit.edu (M.G.)

Received: 26 March 2020; Accepted: 27 March 2020; Published: 31 March 2020

Three-dimensional (3D) bioprinting uses additive manufacturing techniques to fabricate 3D structures consisting of heterogenous selections of living cells, biomaterials, and active biomolecules [1,2]. To date, 3D bioprinting technologies have transformed the fields of tissue engineering and regenerative medicine by enabling fabrication of highly complex biological constructs. Using the patient's medical imaging data, patient- and damage- specific implants can be printed with customized cellular and physiomechanical functionalities [3–5]. The main bioprinting methods include extrusion-based, droplet-based (inkjet), laser-based, and, more recently, vat photopolymerization-based bioprinting [6,7]. A variety of biomaterials (i.e., *bioinks*) have been used for tissue bioprinting, including ceramics, synthetic and natural polymers, decellularized tissues, and more frequently, hybrid bioinks consisting of a combination of these materials [8–11].

While significant and rapid progresses have been made in tissue bioprinting processes for various in vitro applications, such as disease modeling [12] and drug screening [13], there are several challenges to address before bioprinting becomes clinically relevant [14–16]. These constraints include: 1) limited number of available bioink solutions and lack of thorough characterization of their biological and physiomechanical properties [10,17]; 2) poor understanding of the correlation between printed architecture and the ultimate tissue function [18,19]; 3) limitations on the quality of imaging techniques [20,21] and available bioprinters [22]; 4) complex and rather expensive processes involved pre, during, and post-bioprinting [22]; 5) suboptimal, non-specialized printing software and their often incompatibilities [23].

There are eight articles published in this Special Issue composed of four research papers and four review papers. The research articles focus on the influence of electron beam (E-beam) sterilization on in vivo degradation of composite filaments [24], enhancing osteogenic differentiation of stem cells using 3D printed wavy scaffolds [25], the development of a scaffold-free bioprinter [26], and the fabrication of multilayered vascular constructs with a curved structure and multi-branches [27]. Kang et al. investigated the effect of E-beam sterilization on the degradation of β-tricalcium phosphate/polycaprolactone (β-TCP/PCL) composite filaments in a rat subcutaneous model for 24 weeks [24]. Although they reported that the E-beam sterilization accelerated the degradation rate of the composite filaments, due to the decreased crystallinity and decreased molecular weight of PCL after the E-beam irradiation, they concluded that the chemistry of samples plays a bigger role than the sterilization method in biodegradation. Ji and Guvendiren investigated the effect of wavy scaffold architecture on human mesenchymal stem cell (hMSC) osteogenesis by 3D printing as compared to orthogonal scaffold design [25]. They found that when cultured on wavy scaffolds, hMSCs became elongated, formed mature focal adhesions, and showed significantly enhanced osteogenesis. LaBarge et al. developed a custom device enabling the printing of an entire layer of spheroids at once to reduce printing time [26]. They demonstrated the feasibility of this device first using zirconia and alginate beads, which mimic spheroids, and human-induced pluripotent stem cell-derived spheroids. This scaffold-free bioprinter could potentially advance the growing field of scaffold-free 3D bioprinting. Liu et al. developed a combined approached to fabricate multilayered biodegradable vascular constructs for cardiovascular research [27]. In their approach, 3D printing was used to fabricate a mold system which was then used to cast a hydrogel and a sacrificial material. They investigated the channel wall displacement during blood flow using fluid-structure interaction simulations. They also demonstrated the feasibility of their devices using human umbilical vein endothelial cells. Their approach shows a great potential for constructing integrated vasculature for tissue engineering.

The four review articles focused on advanced polymers for 3D organ printing [28], chitosan for tissue and organ bioprinting [29], applications of 3D printing for craniofacial tissue engineering [30], and in vivo tracking of 3D printed tissue-engineered constructs [31]. Wang reviewed advanced polymers exhibiting excellent biocompatibility, biodegradability, 3D printability and structural stability [28]. The author also summarized the challenges of polymers for 3D bioprinting of complex organs. Li et al. reviewed the use of chitosan in tissue repair, including skin, bone, cartilage, and liver tissue, and 3D bioprinting of organs [29]. Tao et al. focused on the applications of 3D printing for craniofacial tissue engineering, including periodontal complex, dental pulp, alveolar bone, and cartilage [30]. Gil et al. reviewed the currently utilized imaging techniques to track tissue engineering scaffolds in vivo, with particular focus on the in vivo tracking of 3D bioprinted tissue constructs [31].

We would like to take this opportunity to express our gratitude to all authors who contributed to this Special Issue. We also wish to thank all the reviewers for dedicating their time to provide thorough and timely reviews to ensure the quality of this Special Issue.

Conflicts of Interest: The authors declare no conflict of interest.

References

1. Cui, H.; Nowicki, M.; Fisher, J.P.; Zhang, L.G. 3D Bioprinting for Organ Regeneration. *Adv. Healthc. Mater.* **2017**, *6*, 1601118. [CrossRef]
2. Murphy, S.V.; Atala, A. 3D bioprinting of tissues and organs. *Nat. Biotechnol.* **2014**, *32*, 773–785. [CrossRef] [PubMed]
3. Heller, M.; Bauer, H.K.; Goetze, E.; Gielisch, M.; Roth, K.E.; Drees, P.; Maier, G.S.; Dorweiler, B.; Ghazy, A.; Neufurth, M.; et al. Applications of patient-specific 3D printing in medicine. *Int. J. Comput. Dent.* **2016**, *19*, 323–339. [PubMed]
4. Luenam, S.; Kosiyatrakul, A.; Hansudewechakul, C.; Phakdeewisetkul, K.; Lohwongwatana, B.; Puncreobutr, C. The Patient-Specific Implant Created with 3D Printing Technology in Treatment of the Irreparable Radial Head in Chronic Persistent Elbow Instability. *Case Rep. Orthop.* **2018**, *2018*, 9272075. [CrossRef] [PubMed]
5. Tomov, M.L.; Cetnar, A.; Do, K.; Bauser-Heaton, H.; Serpooshan, V. Patient-Specific 3-Dimensional-Bioprinted Model for In Vitro Analysis and Treatment Planning of Pulmonary Artery Atresia in Tetralogy of Fallot and Major Aortopulmonary Collateral Arteries. *J. Am. Heart Assoc.* **2019**, *8*, e014490. [CrossRef]
6. Papaioannou, T.G.; Manolesou, D.; Dimakakos, E.; Tsoucalas, G.; Vavuranakis, M.; Tousoulis, D. 3D Bioprinting Methods and Techniques: Applications on Artificial Blood Vessel Fabrication. *Acta Cardiol. Sin.* **2019**, *35*, 284–289. [CrossRef]
7. Li, J.; Chen, M.; Fan, X.; Zhou, H. Recent advances in bioprinting techniques: Approaches, applications and future prospects. *J. Transl. Med.* **2016**, *14*, 271. [CrossRef]
8. Tomov, M.L.; Theus, A.; Sarasani, R.; Chen, H.; Serpooshan, V. 3D Bioprinting of Cardiovascular Tissue Constructs: Cardiac Bioinks. In *Cardiovascular Regenerative Medicine: Tissue Engineering and Clinical Applications*; Serpooshan, V., Wu, S.M., Eds.; Springer International Publishing: Cham, Switzerland, 2019; pp. 63–77. [CrossRef]

9. Gopinathan, J.; Noh, I. Recent trends in bioinks for 3D printing. *Biomater. Res.* **2018**, *22*, 11. [CrossRef]
10. Gungor-Ozkerim, P.S.; Inci, I.; Zhang, Y.S.; Khademhosseini, A.; Dokmeci, M.R. Bioinks for 3D bioprinting: An overview. *Biomater. Sci.* **2018**, *6*, 915–946. [CrossRef]
11. Ji, S.; Guvendiren, M. Recent Advances in Bioink Design for 3D Bioprinting of Tissues and Organs. *Front. Bioeng. Biotechnol.* **2017**, *5*, 23. [CrossRef]
12. Ma, X.; Liu, J.; Zhu, W.; Tang, M.; Lawrence, N.; Yu, C.; Gou, M.; Chen, S. 3D bioprinting of functional tissue models for personalized drug screening and in vitro disease modeling. *Adv. Drug Deliv. Rev.* **2018**, *132*, 235–251. [CrossRef] [PubMed]
13. Mazzocchi, A.; Soker, S.; Skardal, A. 3D bioprinting for high-throughput screening: Drug screening, disease modeling, and precision medicine applications. *Appl. Phys. Rev.* **2019**, *6*, 011302. [CrossRef]
14. Bishop, E.S.; Mostafa, S.; Pakvasa, M.; Luu, H.H.; Lee, M.J.; Wolf, J.M.; Ameer, G.A.; He, T.C.; Reid, R.R. 3-D bioprinting technologies in tissue engineering and regenerative medicine: Current and future trends. *Genes Dis.* **2017**, *4*, 185–195. [CrossRef] [PubMed]
15. Mason, J.; Visintini, S.; Quay, T. An Overview of Clinical Applications of 3-D Printing and Bioprinting. In *CADTH Issues in Emerging Health Technologies*; Canadian Agency for Drugs and Technologies in Health: Ottawa, ON, Canada, 2016; pp. 1–19.
16. Cetnar, A.; Tomov, M.; Theus, A.; Lima, B.; Vaidya, A.; Serpooshan, V. 3D Bioprinting in Clinical Cardiovascular Medicine. In *3D Bioprinting in Medicine: Technologies, Bioinks, and Applications*; Guvendiren, M., Ed.; Springer International Publishing: Cham, Switzerland, 2019; pp. 149–162. [CrossRef]
17. Hu, J.B.; Tomov, M.L.; Buikema, J.W.; Chen, C.; Mahmoudi, M.; Wu, S.M.; Serpooshan, V. Cardiovascular tissue bioprinting: Physical and chemical processes. *Appl. Phys. Rev.* **2018**, *5*, 041106. [CrossRef]
18. Zadpoor, A.A.; Malda, J. Additive Manufacturing of Biomaterials, Tissues, and Organs. *Ann. Biomed. Eng.* **2017**, *45*, 1–11. [CrossRef]
19. Kelly, C.N.; Miller, A.T.; Hollister, S.J.; Guldberg, R.E.; Gall, K. Design and Structure-Function Characterization of 3D Printed Synthetic Porous Biomaterials for Tissue Engineering. *Adv. Healthc. Mater.* **2018**, *7*, e1701095. [CrossRef]
20. Squelch, A. 3D printing and medical imaging. *J. Med. Radiat. Sci.* **2018**, *65*, 171–172. [CrossRef]
21. Rengier, F.; Mehndiratta, A.; von Tengg-Kobligk, H.; Zechmann, C.M.; Unterhinninghofen, R.; Kauczor, H.U.; Giesel, F.L. 3D printing based on imaging data: Review of medical applications. *Int. J. Comput. Assist. Radiol. Surg.* **2010**, *5*, 335–341. [CrossRef]
22. Murphy, S.V.; De Coppi, P.; Atala, A. Opportunities and challenges of translational 3D bioprinting. *Nat. Biomed. Eng.* **2019**, 1–11. [CrossRef]
23. Kacarevic, Z.P.; Rider, P.M.; Alkildani, S.; Retnasingh, S.; Smeets, R.; Jung, O.; Ivanisevic, Z.; Barbeck, M. An Introduction to 3D Bioprinting: Possibilities, Challenges and Future Aspects. *Materials (Basel)* **2018**, *11*, 2199. [CrossRef]
24. Kang, J.-H.; Kaneda, J.; Jang, J.-G.; Sakthiabirami, K.; Lui, E.; Kim, C.; Wang, A.; Park, S.-W.; Yang, P.Y. The Influence of Electron Beam Sterilization on In Vivo Degradation of β-TCP/PCL of Different Composite Ratios for Bone Tissue Engineering. *Micromachines* **2020**, *11*, 273. [CrossRef] [PubMed]
25. Ji, S.; Guvendiren, M. 3D Printed Wavy Scaffolds Enhance Mesenchymal Stem Cell Osteogenesis. *Micromachines* **2019**, *11*, 31. [CrossRef] [PubMed]
26. LaBarge, W.; Morales, A.; Pretorius, D.; Kahn-Krell, M.A.; Kannappan, R.; Zhang, J. Scaffold-Free Bioprinter Utilizing Layer-By-Layer Printing of Cellular Spheroids. *Micromachines* **2019**, *10*, 570. [CrossRef] [PubMed]
27. Liu, Y.; Zhang, Y.; Jiang, W.; Peng, Y.; Luo, J.; Xie, S.; Zhong, S.; Pu, H.; Liu, N.; Yue, T. A Novel Biodegradable Multilayered Bioengineered Vascular Construct with a Curved Structure and Multi-Branches. *Micromachines* **2019**, *10*, 275. [CrossRef]
28. Wang, X. Advanced Polymers for Three-Dimensional (3D) Organ Bioprinting. *Micromachines* **2019**, *10*, 814. [CrossRef]
29. Li, S.; Tian, X.; Fan, J.; Tong, H.; Ao, Q.; Wang, X. Chitosans for Tissue Repair and Organ Three-Dimensional (3D) Bioprinting. *Micromachines* **2019**, *10*, 765. [CrossRef]

30. Tao, O.; Kort-Mascort, J.; Lin, Y.; Pham, M.H.; Charbonneau, M.A.; ElKashty, A.O.; Kinsella, M.J.; Tran, D.S. The Applications of 3D Printing for Craniofacial Tissue Engineering. *Micromachines* **2019**, *10*, 480. [CrossRef]
31. Gil, J.C.; Tomov, L.M.; Theus, S.A.; Cetnar, A.; Mahmoudi, M.; Serpooshan, V. In Vivo Tracking of Tissue Engineered Constructs. *Micromachines* **2019**, *10*, 474. [CrossRef]

Article

The Influence of Electron Beam Sterilization on In Vivo Degradation of β-TCP/PCL of Different Composite Ratios for Bone Tissue Engineering

Jin-Ho Kang [1,†], Janelle Kaneda [2,†], Jae-Gon Jang [1], Kumaresan Sakthiabirami [1], Elaine Lui [3], Carolyn Kim [3], Aijun Wang [4,5,6], Sang-Won Park [1,*] and Yunzhi Peter Yang [2,7,8,*]

1 Department of Prosthodontics, School of Dentistry, Chonnam National University, Gwanju 61186, Korea; jhk.bme1002@gmail.com (J.-H.K.); jangjaegon@naver.com (J.-G.J.); sakthikarthi.dentist@gmail.com (K.S.)
2 Department of Bioengineering, Stanford University, Stanford, CA 94305, USA; jkaneda@stanford.edu
3 Department of Mechanical Engineering, Stanford University, Stanford, CA 94305, USA; elainelui@stanford.edu (E.L.); ck2514@stanford.edu (C.K.)
4 Surgical Bioengineering Laboratory, Department of Surgery, School of Medicine, University of California–Davis, Sacramento, CA 95817, USA; aawang@ucdavis.edu
5 Department of Biomedical Engineering, University of California–Davis, Davis, CA 95616, USA
6 Institute for Pediatric Regenerative Medicine, Shriners Hospitals for Children–Northern California, Sacramento, CA 95817, USA
7 Department of Orthopaedic Surgery, Stanford University, Stanford, CA 94305, USA
8 Department of Materials Science and Engineering, Stanford University, Stanford, CA 94305, USA
* Correspondence: psw320@chonnam.ac.kr (S.-W.P.); ypyang@stanford.edu (Y.P.Y.)
† These individuals are co-first authors.

Received: 13 February 2020; Accepted: 4 March 2020; Published: 6 March 2020

Abstract: We evaluated the effect of electron beam (E-beam) sterilization (25 kGy, ISO 11137) on the degradation of β-tricalcium phosphate/polycaprolactone (β-TCP/PCL) composite filaments of various ratios (0:100, 20:80, 40:60, and 60:40 TCP:PCL by mass) in a rat subcutaneous model for 24 weeks. Volumes of the samples before implantation and after explantation were measured using micro-computed tomography (micro-CT). The filament volume changes before sacrifice were also measured using a live micro-CT. In our micro-CT analyses, there was no significant difference in volume change between the E-beam treated groups and non-E-beam treated groups of the same β-TCP to PCL ratios, except for the 0% β-TCP group. However, the average volume reduction differences between the E-beam and non-E-beam groups in the same-ratio samples were 0.76% (0% TCP), 3.30% (20% TCP), 4.65% (40% TCP), and 3.67% (60% TCP). The E-beam samples generally had more volume reduction in all experimental groups. Therefore, E-beam treatment may accelerate degradation. In our live micro-CT analyses, most volume reduction arose in the first four weeks after implantation and slowed between 4 and 20 weeks in all groups. E-beam groups showed greater volume reduction at every time point, which is consistent with the results by micro-CT analysis. Histology results suggest the biocompatibility of TCP/PCL composite filaments.

Keywords: 3D printing; β-tricalcium phosphate/polycaprolactone (β-TCP/PCL) composite; bone tissue engineering; electron beam sterilization

1. Introduction

There have been many developments in engineering biocompatible and biodegradable bone implant scaffolds for use as an alternative to autografts and allografts for bone defects over the last couple of decades, especially when bone defects are large and donor morbidity is a risk [1–4]. Bone scaffolds have been extensively studied to promote native bone tissue growth and surrounding cell proliferation by optimizing nutrient transportation and mimicking native mechanical properties while minimizing damage to the surrounding tissues [1,2,5–7]. Bone scaffolds have been constructed using various materials such as metals, bioglasses, ceramics, and polymers, and are typically fabricated from a composite of the latter two [1,2,5,7]. Specifically, β-tricalcium phosphate (β-TCP), a polymorph of tricalcium phosphate and a biomimetic ceramic, and polycaprolactone (PCL), a biocompatible polymer, are two commonly used, clinically available biodegradable materials in bone scaffold engineering. TCP has a comparable resorption rate to bone regeneration [1,8,9]. Additionally, when compared to five other commonly used FDA-approved poly(α-hydroxy esters), PCL was one of two that demonstrated the best structural integrity and cellular response [10].

The composite construct of β-TCP and PCL combines the respective benefits of each: osteoconductivity—or bone growth on a surface such as an implant scaffold [11–14]—and easy handling, both of which have only begun to be explored in further depth. The β-TCP/PCL composite is composed of the osteoconductive β-TCP ceramic particles suspended in the bioresorbable PCL polymer matrix [15,16]. The composite material's ability to be extruded into a filament and then 3D-printed enables the creation of controlled, patient-specific scaffolds to optimize its integration within and support for native bone tissue regeneration [17]. In addition to 3D printing, other scaffold fabrication methods include electrospinning, solvent casting, particle leaching, thermally-induced phase separation, and various molding techniques [5,6,18].

Although various factors for optimizing bone scaffolds have been studied, examining the degradation profiles of these constructs is particularly crucial for evaluating the success of bone implants for clinical applications. A bone scaffold should subsist long enough to induce the maximum therapeutic effect at the bone defect site, but also degrade when healing is underway. Poly (α-hydroxy esters)—and by association, composites with polymers in this group—undergo hydrolytic degradation via two methods: surface or bulk [19]. Ideally, degradation and resorption times for bone scaffolds should match bone regeneration rates, depending on the bone defect size. For large bone defects, the degradation and resorption duration for bone scaffolds can be greater than two years [20]. Slow-degrading scaffolds have been shown to prevent tears, allow a slow reintegration of movement, and minimize toxicity at the site of interest when compared to fast-degrading scaffolds [21].

Since sterilization is necessary for the clinical realization of a bone scaffold, it is important to then study how sterilization may change degradation, which further affects the structural integrity and mechanical profiles of bone scaffolds. Various sterilization methods exist for bone scaffolds, including heat-based ethylene oxide immersion and irradiation via ultraviolet, gamma, and electron beam (E-beam) irradiation [22]. Submersion in solvents, such as 70% ethanol, has also been used to sterilize scaffolds, but is insufficient as a sterilization method alone because ethanol has minimal sterilizing power over bacterial spores [22]. Out of all of these methods, E-beam is the most optimal for pre-packaged biomaterials with low melting points, which is relevant for β-TCP/PCL scaffolds [22]. Additionally, E-beam has higher dosage rates than both ultraviolet and gamma irradiation methods, resulting in less exposure time [22]. This is particularly important for polymers like PCL, because irradiation methods like E-beam and gamma have been shown to increase the polydispersity of PCL chains and affect mechanical properties and degradation rates [22–24]. This is a result of PCL ester–ester chain scissioning, in addition to crosslinking, or the formation of chemical bonds to connect polymer chains [23,24].

In our previous study, we found a 14% increase in the initial Young's modulus and a 25% faster in vitro degradation profile for scaffolds that received E-beam compared to those that did not [23]. The increased Young's modulus values after E-beam were likely due to crosslinking, which strengthens the β-TCP/PCL composite structure, while the increase in degradation rate after E-beam in vitro was likely due to chain scissioning, which is thought to weaken the composite structure [23]. Furthermore, since β-TCP particles are merely suspended in the polymer matrix, degradation of β-TCP/PCL scaffolds in any given solution is mainly driven by polymer degradation via the hydrolytic cleavage or scissioning of ester–ester linkages [19,20,23,25]. Previous studies, including ours, have focused solely on 20% TCP/80% PCL [19,20,23,26,27], so this study extends the work by examining the in vivo degradation profiles of various β-TCP/PCL composite ratios by mass (0:100, 20:80, 40:60, and 60:40) in a rat model, particularly studying the effect of E-beam sterilization among these different ratios on in vivo degradation. We have chosen to use extruded filament samples over scaffold samples for this in vivo study for simplification and as a screening test for chemical compositions. While we recognize that scaffolds confer additional properties, such as porosity, that can also influence degradation, the main purpose of our in vivo study is to test how the chemical composition and E-beam affect degradation. This can be achieved using extruded filament samples, while also saving time and cost. In addition, these extruded filaments can help predict the degradation of extrusion-based printed devices and grafts.

2. Materials and Methods

2.1. Sample Fabrication

The sample fabrication protocol was adapted from Bruyas et al. [28]. Four ratios of β-TCP to PCL were synthesized from the stock constituents using a protocol involving dissolution and precipitation phases. The gram-to-gram ratio of β-TCP powder with an average particle size of 100 nm (Berkeley Advanced Biomaterials Inc., Berkeley, CA, USA) to PCL pellets (Sigma-Aldrich, St. Louis, MO, USA) was 0:37.5, 7.5:30, 15:22.5, and 22.5:15 for β-TCP to PCL ratios of 0:100, 20:80, 40:60, and 60:40 by mass, respectively. Materials were suspended in dimethylformamide (DMF) (Fisher Chemical, Waltham, MA, USA): 20 mL DMF per 1 g β-TCP, and 10 mL DMF per 1 g PCL. The materials were gradually mixed into DMF separately by heating each beaker to 70–90 °C and stirring for three hours, before the two were combined and stirred for an additional hour. The mixture was then precipitated into a large container of cold tap water, flattened into a sheet of approximately 200–350 cm^2 area, and dried at room temperature overnight. The composite material was then hand-processed into pellets with diameters of approximately 5 mm. These pellets were fed into a lab-built screw extruder to create a filament with an average diameter of approximately 2.5 mm. Ratios with higher β-TCP content required higher temperatures for extrusion, since β-TCP has a much higher melting point than PCL (1670 °C versus 60 °C, respectively). A 90 °C temperature was used for 0:100 and 20:80, while 100 °C was used for 40:60 and 120 °C for 60:40. This material- and filament-synthesis process was repeated for each of the four ratios. Samples 5 mm long were cut from each filament material for the in vivo study.

2.1.1. Pre-E-Beam Surface Treatment

After fabrication, a pre-E-beam surface treatment (adapted from Bruyas et al. [23]) was administered to all samples to make their surfaces more hydrophilic and rough, which facilitate better degradation solution penetration [29]. Samples were fully immersed in a 5 M NaOH (Ricca Chemical, Arlington, TX, USA) solution from diluting a 10 M stock with purified water (Milli-Q, MilliporeSigma, Burlington, MA, USA) at room temperature for 1 h (40% and 60% β-TCP) and 6.5 h (0% and 20% β-TCP). After NaOH submersion, all samples were rinsed twice with Milli-Q water and dried overnight. Under a sterile biological hood, the samples were then immersed in 70% ethanol for 20 min for sterilization, and then rinsed with PBS (pH 7.4, Gibco, Carlsbad, CA, USA) three times. After drying overnight, the samples were packaged in autoclaved self-sealing sterilization pouches under the sterile biological hood.

2.1.2. E-Beam Specification

Half of all the samples were E-beam irradiated with a standard single dose of 25kGy (Steri-Tek, Fremont, CA, USA), in alignment with the ISO 11137-2:2006 norm. Steri-Tek uses two 10 MeV, 20 KW linear accelerators (Mevex, Stittsville, ON, Canada) to create a DualBeam™ processing method, which increases efficiency by administering uniform doses to products without having to rotate them. The Bruyas et al. study on E-beam and β-TCP/PCL scaffolds also used this E-beam specification [23]. This standard complies with the sterility assurance level (SAL) being less than 10^{-6}. In other words, there can be at most one unsterile item for every one million objects, whether it be devices or scaffolds, in order to qualify as sterile [22].

2.2. The Subcutaneous Implantation of Samples into Rats

For this study, five Sprague Dawley rats (S.D Rat, Taconic Biosciences, Rensselaer, NY, USA) were grown in a pathogen-free environment for a period of 9 weeks. All experiments were conducted in accordance with animal testing ethics and were approved by Chonnam National University Institutional Animal Care and Use Committee (No. CNU IACUC-YB-2018-80). Specimens prepared for in vivo testing were classified as shown in Table 1, and a total of 40 specimens were prepared and divided into eight groups. The rats were anesthetized using 10 mg/kg of Xylazine (Rumpoon, Bayer, Leverkusen, Germany) and 20 mg/kg of Zoletil (Zolazepam + Tiletamine, Virbac, Carros, France) by intraperitoneal injection. To prevent bradycardia, 0.1 mg/kg of an anticholinergic drug (Atropine, JEIL Pharmaceutical, Seoul, Korea) was injected intramuscularly. Both the neck and hind limbs were shaved followed by iodine cure, ethanol (70% ethyl alcohol) disinfection, and incisions. Each filament from the experimental groups was implanted into the neck and hind limbs. Each rat was implanted with eight different groups of cylindrical filaments that were placed in subcutaneous sacs internally and sutured (Vicryl-4.0, Johnson & Johnson Medical, New Brunswick, NJ, USA), with sutures at appropriate intervals to prevent movement of the samples. The transplanted samples were not in contact with each other (Figure 1).

Table 1. Classification codes for each group.

Filament Group		Code	Quantity
E-beam	100% PCL	0, e	5
	20% TCP/80% PCL	20, e	5
	40% TCP/60% PCL	40, e	5
	60% TCP/40% PCL	60, e	5
Non-E-beam	100% PCL	0, no e	5
	20% TCP/80% PCL	20, no e	5
	40% TCP/60% PCL	40, no e	5
	60% TCP/40% PCL	60, no e	5

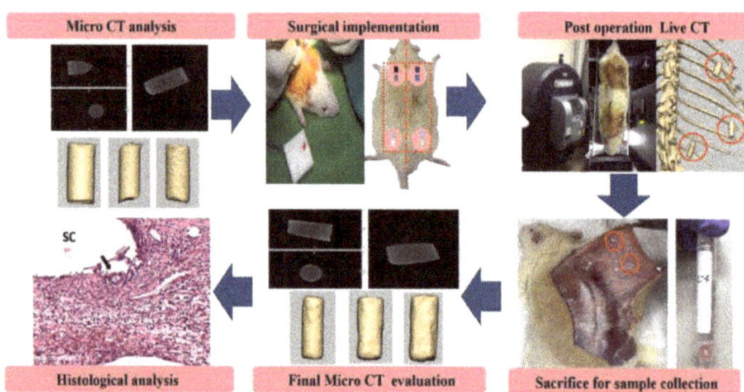

Figure 1. Schematic diagram of the filament implantation and experimental process.

Post-operatively, all the animals received 5 mg/kg of antibiotics (Enrofloxacin, Bayer Leverkusen, Germany) and 5 mg/kg of analgesic anti-inflammatory drugs (Ketoprofen, EagleVet, Seoul, Korea). After implantation, rats were subjected to in vivo live micro-computer tomography (CT) (live-CT) and micro-CT (Figure 1).

2.3. Micro-CT and In Vivo Live-CT

Volumes of the samples before implantation and explants after animal euthanization were measured using micro-CT (SKYSCAN 1272, Bruker, Billerica, MA, USA) at 30 kV voltage, 150 μA current, and 10 μm pixel size. The scanned slices were reconstructed into DICOM files using the Cone Beam program (PerkinElmer, Waltham, MA, USA). After 24 weeks of implantation, the rats were sacrificed, and the implants were removed and subjected to micro-CT measurement under the same conditions as before implantation. The volume change of the specimens was calculated using Equation (1) and an image processing software (Mimics software, Materialize NV, Leuven, Belgium).

$$\text{Micro-CT volume change (\%)} = ((M_{\triangle} - M_0)/M_0) \times 100, \quad (1)$$

where M_0 is the micro-CT data of implants before implantation, and $M_{\triangle}$ is the micro-CT data of explants at 24 weeks after implantation.

The volume changes of the filaments implanted in vivo were measured using a live-CT device (Quantum GX2, PerkinElmer, Inc., USA). Exposure conditions were maintained at 90 kV voltage, 88 μA current, and 90 μm voxel size for 4 min. The volume of the samples was measured after 1 day of implantation and again at 4, 12, and 20 weeks after implantation. The volume change of the specimens was calculated by Equation (2) using an image processing software.

$$\text{Live-CT volume change (\%)} = ((L_{\triangle} - L_0) / L_0) \times 100, \quad (2)$$

where L_0 is the live-CT data at 1 day after implantation, and $L_{\triangle}$ is the data for each set period.

Equation (3) was used to compare the difference between micro-CT (M_{CT}) volume before implantation and live-CT (L_{CT}) volume after 1 day of implantation.

$$\text{Difference between } L_{CT} \text{ and } M_{CT} \text{ (\%)} = ((L_{CT} - M_{CT}))/M_{CT}) \times 100. \quad (3)$$

2.4. Histological Examination

Specimens were fixed using 4% paraformaldehyde solution and demineralized using 10% ethylenediaminetetraacetic acid (EDTA, Sigma-Aldrich, USA). The demineralized specimen was then dehydrated with increasing ethanol concentration (70% to 100%). Subsequently, it was cleaned with Xylene (Sigma-Aldrich, USA), embedded in paraffin, and then cut into 3 μm specimen slices using an Automated Rotary Microtome (Leica RM2255, Leica Microsystems, Wetzlar, Germany). Following the above process, the hydration step was executed. H&E staining (hematoxylin and eosin, Sigma-Aldrich, USA) was performed to evaluate histological characteristics including inflammatory response, collagen presence, and neovascularization.

2.5. Statistics

Quantitative data are presented as mean ± standard deviation with variance analysis according to the Mann–Whitney U test. The Kruskal–Wallis test was used to compare among experimental groups using the PASW Statistics 18.0 software (SPSS Inc., Chicago, IL, USA). $p < 0.05$ was considered statistically significant.

3. Results and Discussion

3.1. Evaluation of Filament Degradation in the Subcutaneous

Figure 2a shows the volume changes of the eight different groups of filaments before and after 24 weeks implantation by micro-CT analysis. The volume change of filaments was used as an indicator of degradation. At 24 weeks, the volume change of groups (0, e) and (0, no e) was 1.54% ± 0.28% and 0.78% ± 0.24%, respectively. Groups (20, e) and (20, no e) was 10.16% ± 3.95% and 6.58% ± 1.04%, respectively. Groups (40, e) and (40, no e) was 10.33 ± 4.19 and 5.68 ± 3.17, respectively. Groups (60, e) and (60, no e) was 10.47 ± 3.59 and 6.80 ± 1.54, respectively. There was a significant difference in volume change between the (0, e) and (0, no e) groups, but there was no significant difference among the other groups because of relatively high standard deviations.

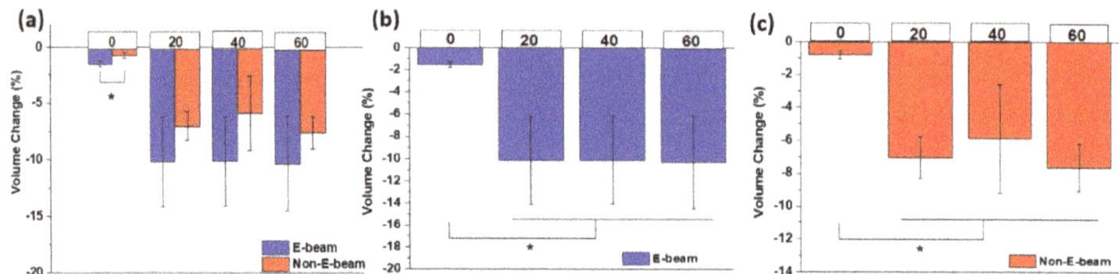

Figure 2. Volume change of filaments by micro-computer tomography (CT) and Equation (1). (**a**) Between E-beam and non-E-beam at 24 weeks (* $p < 0.05$); (**b**,**c**) of 4 different tricalcium phosphate/polycaprolactone (TCP/PCL) ratio groups at 24 weeks (* $p < 0.05$).

However, the volume changes of E-beam groups were greater than the non-E-beam groups in the same TCP:PCL ratio pair. The average difference between the E-beam and non-E-beam groups in each ratio was 0.76% (0% TCP), 3.30% (20% TCP), 4.65% (40% TCP), and 3.67% (60% TCP). The results suggest that E-beam sterilization accelerated degradation, which is consistent with our in vitro degradation study [23] and other literature [19,20,23,26,27]. E-beam accelerates degradation because irradiation causes decreased crystallinity and shorter molecular chains of PCL due to chain scissions. The volume changes of pure PCL filaments are smaller than those of TCP/PCL composite filaments, suggesting that the addition of TCP also accelerated degradation. Compared to the slowly degradable hydrophobic PCL, the higher content of hydrophilic TCP ceramic particles increased water absorption from body fluids and accumulated more lipase enzymes onto the surface layer of the filaments [30,31], inducing hydrolysis and increasing PCL surface erosion [32]. It is worth noting that the increased hydrophilicity of the TCP/PCL surface likely facilitates cell adhesion, which also likely contributed to the increased degradation [32].

There were no significant differences in the degradation among the three different TCP content groups, in both E-beam and non-E-beam (Figure 2b,c). The results suggest that the chemistry of composites plays a bigger role in biodegradation than the sterilization method. In particular, E-beam mainly affected the properties of PCL, not TCP. The concentration of TCP particles embedded into the PCL matrix was not sufficient to allow particles to be interconnected. The lack of interconnection then allows water to penetrate into the center of the composite [19].

In this experiment, live-CT was used to observe the volume change of filaments over time without euthanizing the rats. Using live-CT also reduces the number of rats needed for the study and discrepancies due to the individual characteristics of each rat. However, the 0% TCP group was difficult to scan with live-CT due to its low contrast and fine fiber diameter. The ability to scan with live-CT is affected by the signal-to-noise ratio, the phase difference, and the tissue around the artifact [33–36].

Table 2 lists the volume differences of samples that were estimated by Equation (3) by subtracting the volume before implantation by micro-CT from the volume at one day after implantation by live-CT. The average volumes by live-CT were slightly higher than those by micro-CT in all samples. This difference could be due to the filament swelling inside the live body at one day after implantation. A study found that swelling of PCL occurs rapidly within the first 24 h, with a difference of about 10–20% [37]. In another report, the rate of swelling varies with the amount of hydrophilic material mixed with PCL within 24 h, and the resulting volume increase is around 5–10% [38]. Our results in Table 2 are consistent with the values in these reports.

Table 2. Volume difference between micro-CT and live-CT.

Time/Group	20, e	40, e	60, e	20, no e	40, no e	60, no e
Before surgery (micro-CT) (%)	28.05	21.30	27.73	26.13	22.00	26.24
1 day after implantation (live-CT) (%)	28.83	22.64	29.23	26.93	23.74	27.25
Volume change (%)	+2.78	+6.29	+5.4	+3.09	+7.88	+3.84

Figure 3A shows the volume change over time based on live-CT analysis and Equation (2). All TCP-containing groups exhibited a rapid volume reduction during the first four weeks of implantation. From 4 to 20 weeks, all groups showed a slow and gentle downward slope in volume change, and the E-beam groups resorbed more at every time point. It has been reported that the adsorption of bioactive proteins to the surface of biomaterials from serum and body fluid upon implantation affects the rate of degradation [39], because this influences the effects of cellular interactions on body fluids and active protein substances attached to implanted TCP/PCL specimens. The hydrolysis mechanism of PCL occurred simultaneously, which results in rapid degradation [40].

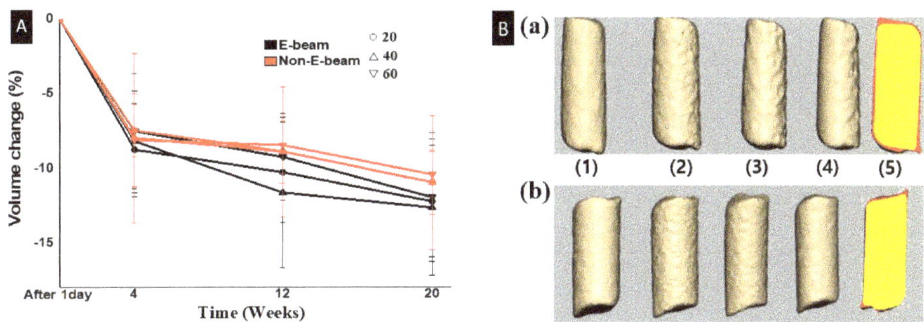

Figure 3. (**A**) Volume change at 1 day, 4, 12, and 20 weeks after implantation by live-CT and Equation (2). (**B**) Three-dimensional model of a sample 40% TCP filament, (**a**) E-beam and (**b**) non-E-beam by live-CT. (1–4) are filaments at 1 day, 4 weeks, 12 weeks, and 20 weeks, respectively. (5) is a superimposition between 1 day (red) and 20 weeks (yellow).

However, one weakness of this study was that we did not use live-CT to examine the volumes of filaments at 24 weeks after implantation, while they were still implanted into the rats; rather, we used micro-CT to examine the volumes of the filaments after explantation. Our decision was misled by our data comparison at day one listed in Table 2, in which the volume by live-CT one day after implantation was greater than the volume by micro-CT before implantation. It turns out that the volumes from micro-CT analysis of the explants after 24 weeks of implantation are actually larger than the values from live-CT analysis at 20 weeks of implantation. The lack of live-CT data at 24 weeks after implantation lost us the opportunity to continuously examine changes between 20 and 24 weeks. As such, we cannot compare the difference between live-CT data before euthanization and micro-CT data after euthanization to further confirm the resolution of CT scanning.

3.2. Histological Examination

Figure 4 shows histology images of sample implants surrounded by tissue. Fibrous encapsulation—a thick and homogeneous colonization by fibroblast cells—was found in all samples.

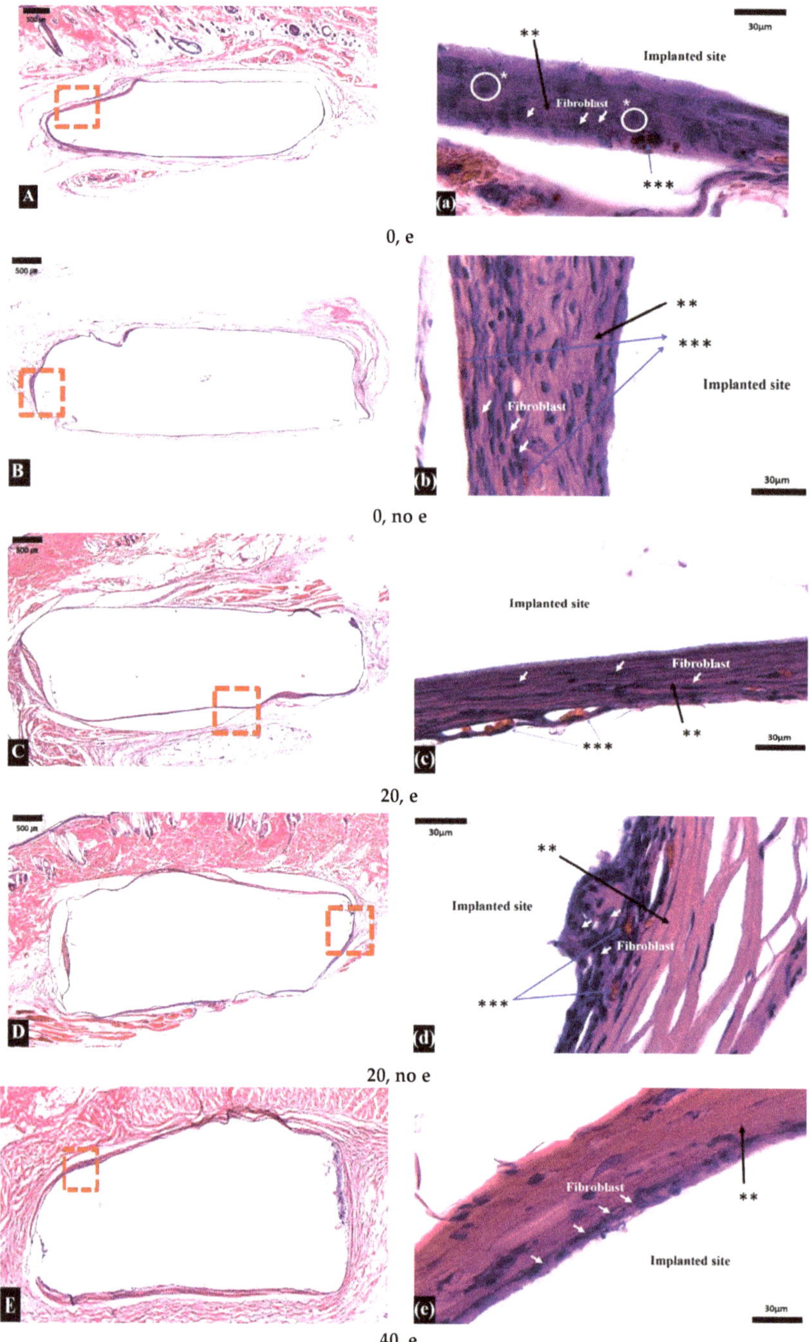

Figure 4. *Cont.*

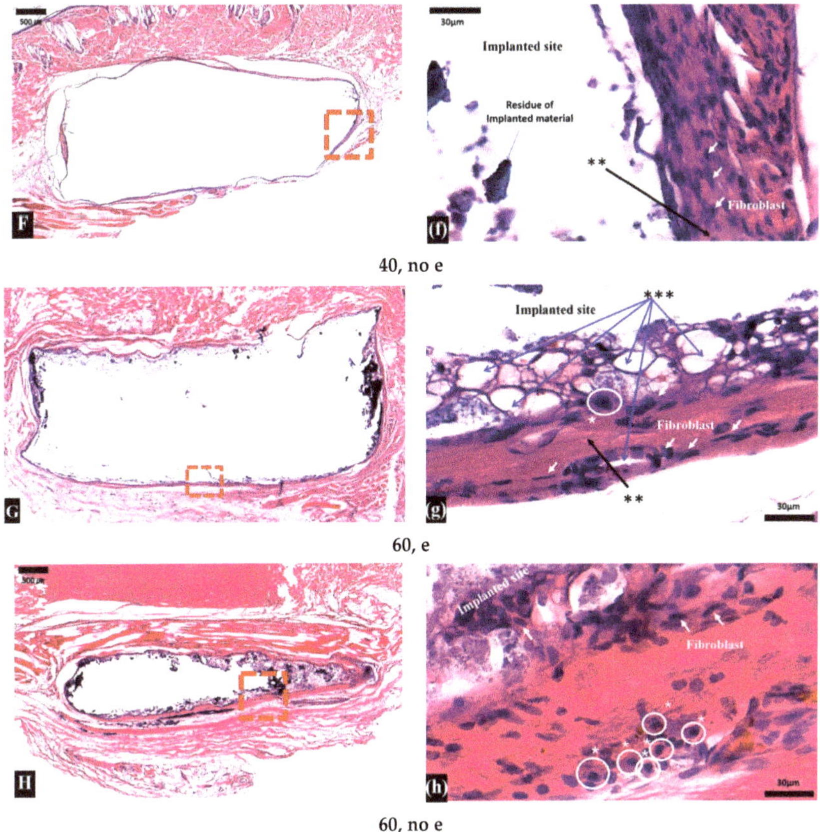

Figure 4. Micrographs of subcutaneous tissue responses to filaments after 24 weeks of implantation. White arrows indicate fibroblasts; (* with a white circle) indicates monocytes and macrophages; (**) indicates protein and collagen; and (***) indicates blood vessels. The micrographic images of each stained samples (10 × and 100 × magnification) are labelled with group code as follow; (**A**) and (**a**) are (0,e), (**B**) and (**b**)are (0,no e), (**C**) and(**c**) are (0,e), (**D**) and(**d**) are(0,no e), (**E**)and (**e**)are(0,e), (**F**)and (**f**)are (0,no e), (**G**) and (**g**)are (0,e),and (**H**) and (**h**)are(0,no e).

Inflammatory macrophages (or monocytes) were found in the (0, e), (0, no e), (60, e), and (60, no e) groups. PCL degradation increases the acidity around the tissues, resulting in an inflammatory response [31]. Fibroblasts were at high-density and accompanied by collagen deposition. Blood vessel formation was also observed. A study of biodegradable polymers by Pêgo et al. shows that encapsulation of samples is mainly composed of macrophages, fibroblasts, and newly formed blood vessels due to immune reactions that are followed by inflammatory reactions, encapsulation characteristics commonly seen after implantation [41]. On the tissue surface of both the E-beam and non-E-beam 60% TCP groups, some multinucleated giant cells (MNGCs) were also observed (data not shown). When macrophages fail to remove foreign bodies due to the presence of slowly degradable PCL, they fuse to form MNGCs, and exhibit foreign body reactions during chronic inflammation [42]. The presence of MNGCs is closely related to the improvement of neovascularization, the degradation and uptake of the implanted biomaterial, and the encapsulation by the transplantation reaction [43,44]. MNGCs invade the implanted biomaterial and begin to destroy the original structure. Blood vessels and connective tissue grow into the biomaterial, leading to premature loss [45].

We understand that subcutaneous implantation may not reflect the degradation of orthotopic bone defect implantation. We used the filaments and subcutaneous implantation as a simple screening test before we test them in bone defects. In our other studies, we implanted 3D-printed TCP/PCL scaffolds of single formulation (20% TCP/80% PCL) into large bone defects in rabbit femoral heads [13,14] and rat femurs [46]. Both implantations showed that our scaffolds promoted bone ingrowth into the porous structure and retained structural integrity, suggesting excellent biocompatibility and osteointegration. The degradation rates of the macroporous 20% TCP/80% PCL scaffolds ranged from 10% to 25% at eight weeks after implantation in femoral head bone defects [13]. The slowest degradation rate in bone defects in rabbits was similar to those of subcutaneous implantation. The difference could be the result of higher surface areas in porous scaffolds versus rods and anatomical sites. We plan to further test E-beam sterilized 3D-printed scaffolds within bone defects in the future.

4. Conclusions

In this study, we extruded β-TCP/PCL filaments of different ratios and sterilized the samples with clinically available E-beam irradiation, and implanted them in subcutaneous sites for 24 weeks. The degradation rates were characterized by micro-CT and live-CT, and biocompatibility of samples was examined by histology analyses. We observed that incorporation of TCP into PCL significantly increased degradation of the composite, but increasing TCP content in the composite did not accelerate degradation. Faster degradation occurred in the first four weeks and gradually slowed down afterward. E-beam sterilization also accelerated degradation, which is due to decreased crystallinity and shorter molecular chains of PCL after the E-beam irradiation. For the TCP/PCL filaments, the chemistry of samples plays a bigger role than the sterilization method in biodegradation. E-beam sterilization did not affect biocompatibility of the implants in the subcutaneous implantation. Our work suggests that creating TCP/PCL composites is a promising method for achieving the degradation properties required to fulfill clinical demands for the use of E-beam sterilization in osteosynthetic applications.

Author Contributions: Conceptualization, Y.P.Y., S.-W.P., J.-H.K., and J.K.; methodology, J.-H.K., J.K., J.-G.J., K.S., E.L., and C.K.; software, J.-H.K., J.K., E.L., and C.K.; validation, J.-H.K., J.K., J.-G.J., K.S., E.L., and C.K.; formal analysis, J.K., J.K., E.L., and C.K.; investigation, Y.P.Y., S.-W.P., J.-H.K., J.K., J.-G.J., K.S., E.L., and C.K.; resources, Y.P.Y. and S.-W.P.; data curation, Y.P.Y., S.-W.P., and A.W.; writing—original draft preparation, J.-H.K., J.K., Y.P.Y., and S.-W.P.; writing—review and editing, Y.P.Y., S.-W.P., E.L., K.S., and A.W.; visualization, J.-H.K. and J.K.; supervision, Y.P.Y. and S.-W.P.; project administration, Y.P.Y. and S.-W.P.; funding acquisition, Y.P.Y. and S.-W.P. All authors have read and agreed to the published version of the manuscript.

Funding: This research was funded through financial support from NIH grants R01AR057837, U01AR069395, R01AR072613; R01AR074458 from NIAMS; DoD grant W81XWH18SBAA1 BA180237; and National Research Foundation of Korea (NRF) grant 2019R1A2C1089456.

Conflicts of Interest: The authors declare no conflict of interest.

References

1. Lichte, P.; Pape, H.C.; Pufe, T.; Kobbe, P.; Fischer, H. Scaffolds for bone healing: Concepts, materials and evidence. *Injury* **2011**, *42*, 569–573. [CrossRef]
2. Stevens, B.; Yang, Y.; Mohandas, A.; Stucker, B.; Nguyen, T.K. A review of materials, fabrication methods, and strategies used to enhance bone regeneration in engineered bone tissues. *J. Biomed. Mater. Res. Part B* **2011**, *85*, 573–582. [CrossRef] [PubMed]
3. Mercado-Pagán, Á.E.; Stahl, A.; Shanjani, Y.; Yang, Y. Vascularization in bone tissue engineering constructs. *Ann. Biomed. Eng.* **2015**, *43*, 718–729. [CrossRef]
4. Yang, Y.; Kang, Y.; Sen, M.; Park, S. Bioceramics in Tissue Engineering. In *Biomaterials for Tissue Engineering Applications*; Burdick, J., Mauck, R., Eds.; Springer: Berlin/Heidelberg, Germany, 2010.
5. Hutmacher, D.W. Scaffolds in tissue engineering bone and cartilage. *Biomaterials* **2000**, *24*, 2529–2543. [CrossRef]
6. Zhao, P.; Gu, H.; Mi, H.; Rao, C.; Fu, J.; Turing, L. Fabrication of scaffolds in tissue engineering: A review. *Mech. Eng.* **2018**, *13*, 107–119. [CrossRef]

7. Winkler, T.; Sass, F.A.; Duda, G.N.; Schmidt-Bleek, K. A review of biomaterials in bone defect healing, remaining shortcomings and future opportunities for bone tissue engineering. *Bone Joint Res.* **2018**, *7*, 232–243. [CrossRef] [PubMed]
8. Giannoudis, P.V.; Dinopoulos, H.; Tsiridis, E. Bone substitutes: An update. *Injury* **2005**, *36*, S20–S27. [CrossRef]
9. McAndrew, M.P.; Gorman, P.W.; Lange, T.A. Tricalcium phosphate as a bone graft substitute in trauma: Preliminary report. *J. Orthop. Trauma* **1988**, *2*, 333–339. [CrossRef]
10. Li, W.; Cooper, J.A.J.r.; Mauck, R.L.; Tuan, R.S. Fabrication and characterization of six electrospun poly (α-hydroxy ester)-based fibrous scaffolds for tissue engineering applications. *Acta Biomater.* **2006**, *2*, 377–385. [CrossRef]
11. Albrektsson, T.; Johansson, C. Osteoinduction, osteoconduction and osteointegration. *Eur. Spine J.* **2001**, *10*, S96–S101. [CrossRef]
12. Shanjani, Y.; Kang, Y.; Zarnescu, L.; Ellerbee Bowden, A.K.; Koh, J.T.; Ker, D.F.E.; Yang, Y. Endothelial pattern formation in hybrid constructs of additive manufactured porous rigid scaffolds and cell-laden hydrogels. *J. Mech. Behav. Biomed.* **2017**, *65*, 356–372. [CrossRef] [PubMed]
13. Kawai, T.; Shanjani, Y.; Fazeli, S.; Behn, A.W.; Okuzu, Y.; Goodman, S.B.; Yang, Y.P. Customized, degradable, functionally graded scaffold for potential treatment of early stage osteonecrosis of the femoral head. *J. Orthop. Res.* **2018**, *36*, 1002–1011. [CrossRef]
14. Maruyama, M.; Nabeshima, A.; Pan, C.C.; Behn, A.W.; Thio, T.; Lin, T.; Pajarinen, J.; Kawai, T.; Takagi, M.; Goodman, S.B.; et al. The effects of a functionally-graded scaffold and bone marrow-derived mononuclear cells on steroid-induced femoral head osteonecrosis. *Biomaterials* **2018**, *187*, 39–46. [CrossRef] [PubMed]
15. Legeros, R.Z. Calcium phosphate-based osteoinductive materials. *Chem. Rev.* **2008**, *108*, 4742–4753. [CrossRef]
16. Woodruff, M.A.; Hutmacher, D.W. The return of a forgotten polymer—Polycaprolactone in the 21st century. *Prog. Polym. Sci.* **2010**, *35*, 1217–1256. [CrossRef]
17. Bose, S.; Vahabzadeh, S.; Bandyopadhyay, A. Bone tissue engineering using 3D printing. *Mater. Today* **2013**, *16*, 496–504. [CrossRef]
18. Derakhshanfar, S.; Mbeleck, R.; Xu, K.; Zhang, X.; Zhong, W.; Xing, M. 3D bioprinting for biomedical devices and tissue engineering: A review of recent trends and advances. *Bioact. Mater.* **2018**, *3*, 144–156. [CrossRef]
19. Lam, C.X.; Hutmacher, D.W.; Schantz, J.T.; Woodruff, M.A. Evaluation of polycaprolactone scaffold degradation for 6 months in vitro and *in vivo*. *J. Biomed. Mater. Res. Part A* **2009**, *90*, 906–919. [CrossRef]
20. Lam, C.X.; Teoh, S.H.; Hutmacher, D.W. Comparison of the degradation of polycaprolactone and polycaprolactone–(b-tricalcium phosphate) scaffolds in alkaline medium. *Polym. Int.* **2007**, *56*, 718–728. [CrossRef]
21. Ker, D.F.E.; Wang, D.; Behn, A.W.; Wang, E.T.H.; Zhang, X.; Zhou, B.Y.; Mercado-Pagán, Á.E.; Kim, S.; Kleimeyer, J.; Gharaibeh, B.; et al. Functionally Graded, Bone- and Tendon-Like Polyurethane for Rotator Cuff Repair. *Adv. Funct. Mater.* **2018**, *28*, 1–16. [CrossRef]
22. Baume, A.S.; Boughton, P.C.; Coleman, N.V.; Ruys, A.J. Sterilization of tissue scaffolds. In *Characterisation and Design of Tissue Scaffolds*, 1st ed.; Tomlins, P., Ed.; Woodhead Publishing: Cambridge, UK, 2016.
23. Bruyas, A.; Moeinzadeh, S.; Kim, S.; Lowenberg, D.W.; Yang, Y.P. Effect of Electron Beam Sterilization on Three-Dimensional-Printed Polycaprolactone/Beta-Tricalcium Phosphate Scaffolds for Bone Tissue Engineering. *Tissue Eng. Part A* **2019**, *25*, 1–9. [CrossRef]
24. Cottam, E.; Hukins, D.W.; Lee, K.; Hewitt, C.; Jenkins, M.J. Effect of sterilisation by gamma irradiation on the ability of polycaprolactone (PCL) to act as a scaffold material. *Med. Eng. Phys.* **2009**, *31*, 221–226. [CrossRef] [PubMed]
25. Díaz, E.; Sandonis, I.; Valle, M.B. In Vitro Degradation of Poly(caprolactone)/nHA Composites. *J. Nanomater.* **2014**, *2014*, 1–8. [CrossRef]
26. Yeo, A.; Rai, B.; Sju, E.; Cheong, J.J.; Teoh, S.H. The degradation profile of novel, bioresorbable PCL-TCP scaffolds: An in vitro and in vivo study. *J. Biomed. Mater. Res. Part. A* **2007**, *84*, 208–218. [CrossRef] [PubMed]
27. Lei, Y.; Rai, B.; Ho, K.H.; Teoh, S.H. In vitro degradation of novel bioactive polycaprolactone—20% tricalcium phosphate composite scaffolds for bone engineering. *Mater. Sci. Eng. C* **2007**, *27*, 293–298. [CrossRef]

28. Bruyas, A.; Lou, F.; Stahl, A.M.; Gardner, M.; Maloney, W.; Goodman, S.; Yang, Y.P. Systemic characterization of 3D-printed PCL/β-TCP scaffolds for biomedical devices and bone tissue engineering: Influence of composition and porosity. *J. Mater. Res.* **2018**, *0*, 1–12. [CrossRef]
29. Schantz, J.T.; Teoh, S.H.; Lim, T.C.; Endres, M.; Lam, C.X.; Hutmacher, D.W. Repair of calvarial defects with customized tissue-engineered bone grafts 1. Evaluation of osteogenesis in a three-dimensional culture system. *Tissue Eng.* **2003**, *9*, S113–S126. [CrossRef]
30. Darwis, D.; Mitomo, H.; Yoshii, F. Degradability of radiation crosslinked PCL in the supercooled state under various environments. *Polym. Degrad. Stab.* **1999**, *65*, 279–285. [CrossRef]
31. Sung, H.J.; Meredith, C.; Johnson, C.; Galis, Z.S. The effect of scaffold degradation rate on three-dimensional cell growth and angiogenesis. *Biomaterials* **2004**, *25*, 5735–5742. [CrossRef]
32. Thuaksuban, N.; Pannak, R.; Boonyaphiphat, P.; Monmaturapoj, N. In vivo biocompatibility and degradation of novel Polycaprolactone-Biphasic Calcium phosphate scaffolds used as a bone substitute. *Biomed. Mater. Eng.* **2018**, *29*, 253–267. [CrossRef]
33. Hsieh, J. *Computed Tomography: Principles, Design, Artifacts, and Recent Advances*, 2nd ed.; SPIE: Bellingham, WA, USA, 2009.
34. Buzug, T.M. Computed Tomography. In *Springer Handbook of Medical Technology*; Kramme, R., Hoffmann, K.P., Pozos, R., Eds.; Springer: Berlin/Heidelberg, Germany, 2011.
35. Wang, Y.; Garcea, S.C.; Withers, P.J. 7.6 Computed tomography of composites. *Compr. Comp. Mater.* **2018**, *7*, 101–118.
36. Davis, G.R.; Elliott, J.C. Artefacts in X-ray microtomography of materials. *Mater. Sci. Technol. Ser.* **2006**, *22*, 1011–1018. [CrossRef]
37. Fadaie, M.; Mirzaei, E. Nanofibrillated chitosan/polycaprolactone bionanocomposite scaffold with improved tensile strength and cellular behavior. *Nanomed. J.* **2019**, *5*, 77–89. [CrossRef]
38. Kim, J.J.; Singh, R.K.; Seo, S.J.; Kim, T.H.; Kim, J.H.; Lee, E.J.; Kim, H.W. Magnetic scaffolds of polycaprolactone with functionalized magnetite nanoparticles: Physicochemical, mechanical, and biological properties effective for bone regeneration. *RSC Adv.* **2014**, *4*, 17325–17336. [CrossRef]
39. Wypych, G. Microscopic Mechanisms of Damage Caused by Degradants. In *Atlas of Material Damage*, 2nd ed.; ChemTec Publishing: Toronto, ON, Canada, 2017.
40. Vohra, S.; Hennessy, K.M.; Sawyer, A.A.; Zhuo, Y.; Bellis, S.L. Comparison of mesenchymal stem cell and osteosarcoma cell adhesion to hydroxyapatite. *J. Mater. Sci. Mater. Med.* **2008**, *19*, 3567. [CrossRef]
41. Pêgo, A.; Van Luyn, M.; Brouwer, L.; Van Wachem, P.; Poot, A.A.; Grijpma, D.W.; Feijen, J. In vivo behavior of poly (1,3-trimethylene carbonate) and copolymers of 1,3-trimethylene carbonate with D,L-lactide or ε-caprolactone: Degradation and tissue response. *J. Biomed. Mater. Res. Part A* **2003**, *67*, 1044–1054. [CrossRef]
42. Anderson, J.M.; Rodriguez, A.; Chang, D.T. Foreign body reaction to biomaterials. *Semin. Immunol.* **2008**, *20*, 86–100. [CrossRef]
43. Barbeck, M.; Kubesch, A.; Booms, P.; Boehm, N.; Choukroun, J.; Sader, R.; Kirkpatrick, C.J.; Lorenz, J.; Ghanaati, S. Porcine dermis-derived collagen membranes induce implantation bed vascularization via multinucleated giant cells: A physiological reaction? *J. Oral Implantol.* **2015**, *41*, 238–251. [CrossRef]
44. Barbeck, M.; Lorenz, J.; Holthaus, M.G.; Raetscho, N.; Kubesch, A.; Booms, P.; Sader, R.; Kirkpatrick, C.J.; Ghanaati, S. Porcine Dermis and Pericardium-Based, Non–Cross-Linked Materials Induce Multinucleated Giant Cells After Their In Vivo Implantation: A Physiological Reaction? *J. Oral Implantol.* **2015**, *41*, 267–281. [CrossRef]
45. Barbeck, M.; Najman, S.; Stojanović, S.; Mitić, Ž.; Živković, J.M.; Choukroun, J.; Kovačević, P.; Sader, R.; Kirkpatrick, C.J.; Ghanaati, S. Addition of blood to a phycogenic bone substitute leads to increased in vivo vascularization. *Biomed. Mater.* **2015**, *10*, 1–15. [CrossRef]
46. DeBaun, M.R.; Stahl, A.M.; Daoud, A.I.; Pan, C.C.; Bishop, J.A.; Gardner, M.J.; Yang, Y.P. Preclinical Induced Membrane Model to Evaluate Synthetic Implants for Healing Critical Bone Defects Without Autograft. *J. Orthop. Res.* **2019**, *37*, 60–68. [CrossRef] [PubMed]

Article

3D Printed Wavy Scaffolds Enhance Mesenchymal Stem Cell Osteogenesis

Shen Ji [1] **and Murat Guvendiren** [1,2,*]

1 Otto H. York Department of Chemical and Materials Engineering, New Jersey Institute of Technology, University Heights, Newark, NJ 07102, USA; SJ422@njit.edu

2 Department of Biomedical Engineering, New Jersey Institute of Technology, University Heights, Newark, NJ 07102, USA

* Correspondence: muratg@njit.edu; Tel.: +1-973-596-2932

Received: 22 November 2019; Accepted: 21 December 2019; Published: 25 December 2019

Abstract: There is a growing interest in developing 3D porous scaffolds with tunable architectures for bone tissue engineering. Surface topography has been shown to control stem cell behavior including differentiation. In this study, we printed 3D porous scaffolds with wavy or linear patterns to investigate the effect of wavy scaffold architecture on human mesenchymal stem cell (hMSC) osteogenesis. Five distinct wavy scaffolds were designed using sinusoidal waveforms with varying wavelengths and amplitudes, and orthogonal scaffolds were designed using linear patterns. We found that hMSCs attached to wavy patterns, spread by taking the shape of the curvatures presented by the wavy patterns, exhibited an elongated shape and mature focal adhesion points, and differentiated into the osteogenic lineage. When compared to orthogonal scaffolds, hMSCs on wavy scaffolds showed significantly enhanced osteogenesis, indicated by higher calcium deposition, alkaline phosphatase activity, and osteocalcin staining. This study aids in the development of 3D scaffolds with novel architectures to direct stem osteogenesis for bone tissue engineering.

Keywords: biomaterials; additive manufacturing; stem cells; tissue engineering; bone regeneration

---

1. Introduction

There is a growing interest in developing porous scaffolds for bone tissue engineering enabling temporary mechanical support for cells to attach, migrate, and produce newly formed extracellular matrix to form functional bone tissue ultimately [1–3]. Although bone has a robust regenerative ability, therapeutic interventions are required for large bone defects [4,5]. Grafts (autografts, allografts, and xenografts) are commonly used in the clinic to fill the defect site and to regenerate bone tissue [6,7]. Porous scaffolds can be considered as an alternative to regenerate bone while mechanically supporting the defect site [8]. A wide range of techniques have been developed to fabricate porous bone scaffolds, such as gas foaming [9–11], solvent casting and particle/salt leaching [12–16], phase separation [17,18], freeze drying [19,20], and electrospinning [21–23]. However, the majority of these techniques fail to control the 3D architecture of the scaffolds precisely, including pore size and pore distribution, and also fail to develop reproducible scaffolds [1]. 3D printing is an additive manufacturing technique and enables the fabrication of custom designed and highly complex 3D scaffolds. 3D printing allows the use of the patient's own medical images to design personalized scaffolds that are anatomically similar to the defect site. Thus, it has been widely utilized for fabricating custom designed bone scaffolds [24–28]. A wide range of 3D printing techniques have been used to fabricate 3D bone scaffolds, such as fused deposition modeling (FDM) [29–32], direct ink writing (DIW) [33,34], selective laser sintering and melting (SLS and SLM) [35], stereolithography (SLA) [36–38], continuous digital light processing (cDLP) [39,40], and inkjet printing [41,42]. These 3D printing technologies allow utilizing various printable materials [43] and designs [44]. Computational tools have also been utilized to optimize scaffold architecture to achieve enhanced permeability and mechanical properties [45–49].

Mesenchymal stem cells (MSCs) are regarded as a clinically relevant cell source for bone tissue engineering due to their ability to proliferate and migrate, as well as their potential to differentiate into the osteogenic lineage (bone) [50–53]. Stem cells are known to feel and respond to their microenvironment by regulating their function [54–57]. Materials based approaches have been developed to engineer extracellular matrix (ECM) mimetic microenvironments [58–60], including macro- and nano-scale topographical cues to control stem cell behavior [61,62]. Topographical cues alone have been shown to control stem cell response, such as morphology, alignment, proliferation, migration, cytoskeletal organization, focal adhesion, nuclear deformation, and differentiation [62–64]. For example, human MSCs (hMSCs) are shown to produce bone mineral when cultured on substrates with the nanoscale order [65]. Nanoscale roughness is shown to enhance MSC osteogenesis even in the absence of induction media [66,67]. This phenomenon is shown to be due to clustering of absorbed proteins on a nano-topography, which promotes integrin mediated focal adhesions, enhancing cellular contractility and stem cell osteogenesis [66]. Microscale patterns confining stem cells within cell adhesive regions were used to control stem cell shape or cellular spreading. For instance, McBeath et al. showed that hMSCs with the spread morphology led to actin-myosin generated tension and promoted osteogenic differentiation [68]. Increasing cellular contractility, or cytoskeletal tension, by changing the shape of the multicellular sheets, Ruiz and Chen were able to enhance osteogenic differentiation of hMSCs [69]. Mrksich and co-workers showed that stem cells residing on curved surfaces became highly contractile and differentiated to the osteogenic lineage [70]. Lineage commitment of hMSCs on hydrogel wrinkling patterns was determined by the pattern morphology, such that hMSCs on lamellar patterns formed a spread morphology with a high cell aspect ratio (>4) differentiated into osteogenic progenitors [71]. When porous 3D scaffolds are considered, pore architecture, surface topography, and interconnectivity are shown to control osteogenic differentiation of human mesenchymal progenitor cells [72]. Simon and co-workers fabricated 2D films and 3D porous scaffolds with different techniques (gas foaming, salt leaching, phase separation, electrospinning, 3D printing, and spin coating) to examine the seeded hMSCs' osteogenesis, which indicated that the scaffolds could be optimized to control the cell morphology to direct differentiation [73]. Recently, DIW was used to create 3D scaffolds with distinct architectures composed of square (SQR), hexagonal (HEX), or octagonal (OCT) patterns [74]. Human MSCs were reported to exhibit a higher cell aspect ratio and mean cell area on OCT scaffolds as compared to SQR and HEX scaffolds, and hence showed significantly enhanced osteogenic differentiation. Although the effect of curvature is well documented in 2D, it has not yet been studied systematically in 3D.

In this work, we used 3D printing to fabricate wavy poly(caprolactone) (PCL) scaffolds to investigate the effect of curvature on hMSC osteogenesis. A sinusoidal waveform was used to create wavy scaffolds. The wavelength and amplitude of the sinusoid were systematically varied to design five distinct wavy scaffolds. An orthogonal scaffold with straight struts was used as a control. First, we investigated the effects of scaffold architecture on stem cell growth, including cell attachment, proliferation, and shape (spreading). Then, we studied the osteogenic differentiation of hMSCs on wavy scaffolds as compared to the commonly used orthogonal architecture. The main hypothesis behind this study was that the wavy scaffolds can direct a more elongated and stretched stem cell morphology, resulting in highly organized cytoskeletal arrangement with high contractility. This could lead to an increased osteogenesis, the degree of which can be controlled by the degree of the curvature or waviness.

2. Materials and Methods

2.1. *Scaffold Design*

Autodesk® Fusion 360™ (Autodesk Inc., San Rafael, CA, USA) was used to design the 3D models. The basic 3D model was designed as a cylinder with a diameter of 15 mm and a height of 1 mm. The 3D model (.stl file) was then loaded into Perfactory RP for slicing, with a layer height equal to 0.25 mm. The sliced file (.bpl file) was loaded into Visual Machine, and the infill patterns were selected. A linear pattern was selected for the orthogonal scaffolds (i.e., the control group), and a sinusoidal waveform was selected for the wavy scaffolds (Figure 1). For wavy scaffolds, the amplitude and the wavelength of the sinusoid were varied systematically to develop 5 distinct scaffold designs (Table 1).

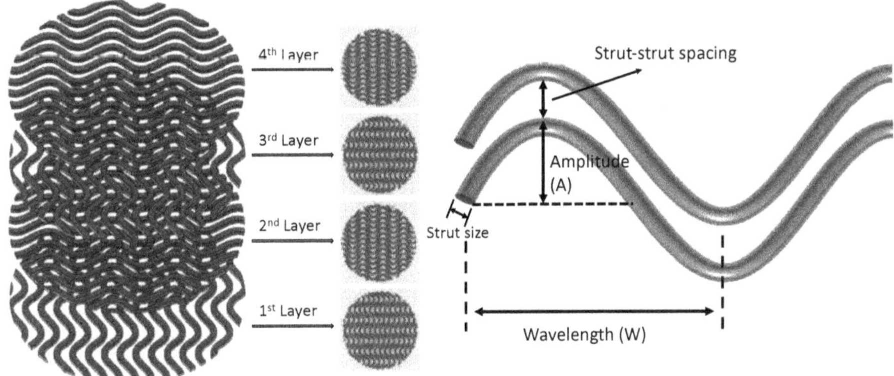

Figure 1. Wavy scaffold design containing 4 layers (**left**) and schematic showing the strut design for wavy scaffolds (**right**).

Table 1. Design and printing parameters for the scaffolds.

Parameter	Orthogonal	A05W2	A0.5W3	A0.5W4	A0.75W4	A1W4
Amplitude (mm)	-	0.5	0.5	0.5	0.75	1
Wavelength (mm)	-	2	3	4	4	4
Strut diameter (μm)	533 ± 9	497 ± 73	490 ± 36	513 ± 30	510 ± 32	460 ± 58
Strut spacing [1] (μm)	395 ± 6	277 ± 59	350 ± 53	396 ± 72	308 ± 72	336 ± 103
Struts per layer	16	15	15	15	15	15
Temperature [2] (°C)	80	80	80	80	80	80
Print pressure (Pa)	7×10^5	7×10^5	7×10^5	7×10^5	7×10^5	7×10^5
Print speed (mm/s)	4	4	6	5	5	5
E [3] (MPa)	12.4 ± 0.3	10.5 ± 0.5	11.5 ± 0.2	9.5 ± 0.2	10.7 ± 0.2	11.3 ± 0.5
Porosity [4] (%)	56.3 ± 0.7	56.5 ± 1.2	55.9 ± 0.3	61.7 ± 0.9	57.6 ± 0.5	57.2 ± 3.1

[1] Strut-to-strut distance. [2] Print temperature. [3] Young's modulus from compression tests. [4] Micro-CT results.

2.2. 3D Printing of Scaffolds

3D Bioplotter (EnvisionTEC, Gladbeck, Germany) was used to print the scaffolds using PCL pellets (MW = 55 kDa, Polysciences Inc., Warrington, PA, USA). The print temperature and pressure were set to 80 °C and 700 kPa (7 bar), whereas the print speed was varied from 4 to 6 mm/s for each design to achieve a similar strut size (see Table 1 for actual values for each design).

2.3. Characterization of the Scaffolds

3D printed scaffolds were imaged by using a scanning electron microscope (SEM, JSM-7900F, JEOL, Tokyo, Japan) and a micro-computed tomography scanner (micro-CT, SkyScan 1275, Bruker, Billerica, MA, USA). SEM images were used to measure the strut size and the strut-to-strut distance. Micro-CT was used to measure the porosity of the scaffolds. Compression tests were performed on 3D printed scaffolds using an Instron machine (model 3343) with a 1000 N load cell and a 0.5 mm/min displacement rate. Three samples for each scaffold group were tested.

2.4. Preparation of the Scaffolds for Cell Culture

Scaffolds were sterilized by immersing them in a 75% ethanol solution for 30 min, followed by 1 h of ultraviolet (UV) light exposure (by a germicidal lamp) for each side of the scaffold. Scaffolds were then incubated in 300 µL of fibronectin solution (20 µg/mL, bovine fibronectin plasma, Invitrogen) overnight to enhance cell attachment. Fibronectin solution was removed, and scaffolds were washed with Dulbecco's Phosphate Buffered Saline (DPBS, Gibco, New York, NY, USA). Scaffolds were then moved into a new well and kept in growth media prior to cell seeding.

2.5. Cell Culture and Reagents

Human mesenchymal stem cells (hMSCs, passage 4, Lonza, Walkersville, MD, USA) were cultured in growth media (α-MEM (minimum essential medium) supplemented with 10% fetal bovine serum (FBS, Gibco, New York, NY, USA) and 1% penicillin-streptomycin (pen-strep, Gibco, New York, NY, USA)). Prior to seeding, each scaffold was removed from the growth media and placed in a single well in a non-treated 24 well plate. The human mesenchymal stem cell (hMSC) suspension (133,000 cells/mL) was seeded from the top of the scaffolds (300 µL per scaffold, corresponding to approximately 5000 cells/cm^2). Cells were incubated at 37 °C for 60 min to allow cell attachment. Scaffolds were then flipped, and the same amount of cell suspension was seeded from the top, followed by 60 min incubation at 37 °C. The scaffolds were then transferred to a new non-treated 24 well plate, and 1 mL of fresh growth media was added into each well. The scaffolds were incubated for 7 days in growth media. For osteogenic differentiation studies, growth media was replaced with osteogenic induction media (hMSC osteogenic differentiation medium BulletKit™, Lonza, Basel, Switzerland) at Day 7, and cells were cultured for an additional 14 days. The media was refreshed every 3 days in cell culture studies.

2.6. Cell Culture and Characterization

For stem cell growth studies, the AlamarBlue assay (AlamarBlue™ Cell Viability Reagent, Invitrogen) and PicoGreen assay (Quant-iT™ PicoGreen™ dsDNA Assay Kit, Invitrogen) were used to evaluate the cell proliferation at Days 1, 4, and 7, according to the manufacturer's protocol. A Tecan plate reader (Infinite M200 Pro, Tecan, Männedorf, Switzerland) was used to complete the assays for these studies. To visualize the attached cells on the scaffolds, cells were washed with DPBS (3×), fixed with 4% formaldehyde for 15 min, followed by DPBS wash (3×), and permeabilization in 0.25% Triton-X DPBS solution for 1 h. Cells were stained for F-actin using rhodamine phalloidin (1:40 in DPBS, Invitrogen). Cell nuclei were stained with 4′,6-diamidino-2-phenylindole (DAPI, 1:2000 in DPBS, Sigma, St. Louis, MO, USA). At Day 7, cells were immunostained for vinculin using the anti-vinculin-FITC antibody (1:50, mouse monoclonal, Sigma). For this purpose, cells were incubated in 10% goat serum (in PBS) for 30 min, washed with staining solution (3×, 3% bovine serum albumin + 0.1% Tween-20 +0.25% Triton-X), and incubated in vinculin antibody in staining solution overnight at 4 °C. Cells were imaged by using a confocal and multiphoton microscopy (TCS SP8 MP, Leica, Wetzlar, Germany).

For differentiation studies, calcium deposition was evaluated at Day 21 by using the alizarin red staining kit (AR, Sigma, St. Louis, MO, USA). After staining was completed, cells were washed with DPBS (3×), and incubated in 10% cetylpyridinium chloride (Sigma, St. Louis, MO, USA) in sodium phosphate buffer (10 mM, pH 7, Sigma) to remove the stain. This solution was then used to quantify calcium content by using a Tecan plate reader (scanned at 405 nm). Alkaline phosphatase activity was studied with the QuantiChrom™ Alkaline Phosphatase Assay Kit (ALP assay Kit, BioAssay Systems, Hayward, NY, USA). Cells cultured within the scaffolds were first lysed with 0.2% Triton-X followed by 3 freeze-thaw circles. The lysate was then reacted with p-nitrophenyl phosphate working solution and scanned at 405 nm using a plate reader (Infinite M200 Pro, Tecan). For osteocalcin (OC) staining, cells were fixed at Day 14 and Day 21. Cells were incubated in 10% goat serum (in PBS) for 30 min, washed with staining solution (3×, 3% bovine serum albumin +0.1% Tween-20 +0.25% Triton-X), and incubated with the OC primary antibody (1:200, monoclonal mouse, Invitrogen) in the staining solution overnight at 4 °C. After removing the antibody containing staining solution and washing the samples with fresh staining solution, cells were incubated in Alexa Fluor 488 rabbit anti-mouse secondary antibody (1:100, Invitrogen) in staining solution for 2 h. Samples were then stained with phalloidin (rhodamine phalloidin, Invitrogen) and DAPI to visualize F-actin and cell nuclei, respectively. Cells were imaged by using a confocal and a multiphoton microscopy (TCS SP8 MP, Leica). All of the collected images were processed using ImageJ (NIH, Bethesda, MD, USA) for further analysis.

2.7. Statistics

The data were analyzed using Origin 2016 software. Data are presented as the mean ± the standard deviation. One way ANOVA with Tukey's HSD post hoc test of means was used to make comparisons between sample groups ($n \geq 3$ samples per group unless otherwise specified).

3. Results

3.1. 3D Printing of PCL Scaffolds

PCL scaffolds with six distinct designs, including one linear design (orthogonal) and five wavy designs in the form of a sinusoidal wave with varying amplitude (A) and wavelength (W) (A0.5W2, A0.5W3, A0.5W4, A0.75W4, and A1W4, where the numbers following A and W denote the actual values of A and W in mm) were printed (Table 1). Figure 2 shows the pictures, micro-CT images, and SEM images of the scaffolds. SEM images were used to measure the printed strut width and spacing between struts for each design, and the results are summarized in Table 1. Briefly, when all the designs were considered, the average strut width was within the range of 460 ± 58 to 533 ± 9 µm, and the spacing between struts (strut-to-strut distance) was within the range of 277 ± 59 to 395 ± 6 µm.

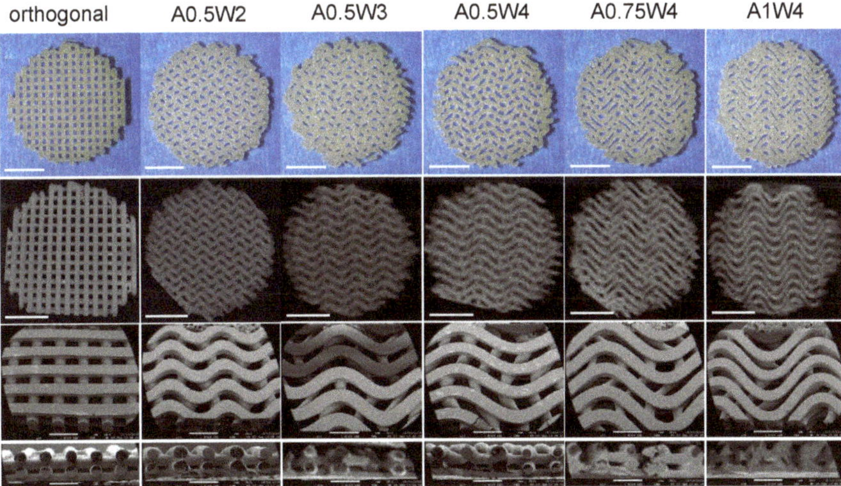

Figure 2. Images of the scaffolds. From the top to bottom row, images correspond to pictures (top view), micro-computed tomography (micro-CT) images (top view), scanning electron microscope (SEM) images, and SEM cross-section images. Scale bars are 500 microns for pictures and micro-CT images and 1 mm for SEM images.

3.2. Mechanical Tests

Compression tests were performed on each sample group, and the results are summarized in Table 1 and Figure 3. The compressive modulus (Young's modulus, E) of all the designs was in the range of 9.5–12.4 MPa (Table 1). E (9.5 MPa) for the A0.5W4 design (with the highest porosity, ~62%) was significantly lower than the rest of the sample groups. The orthogonal design (E = 12.4 MPa and porosity = ~56%) showed significantly higher E as compared to A0.5W2 (E = 10.5 MPa and porosity = ~56%), A0.5W4, and A1W4 (E = 11.3 MPa and porosity = ~57%).

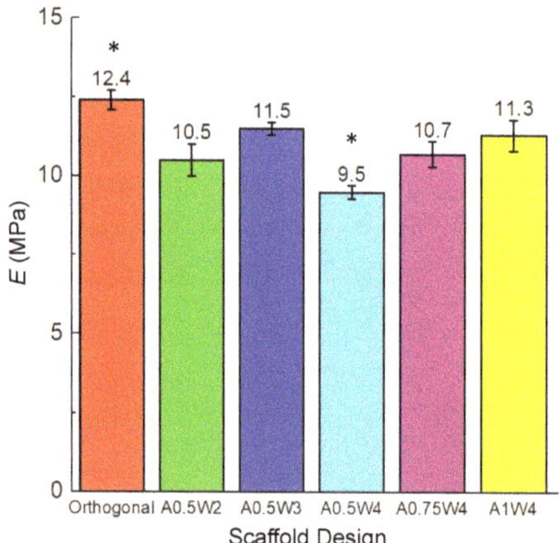

Figure 3. Young's modulus (E) values of the scaffolds for each scaffold design. * $p < 0.005$ for orthogonal vs. A05W2, A0.5W4, and A1W4; and for A0.5W4 vs. all sample groups.

3.3. Growth Study

The hMSC growth studies were performed by culturing cells in growth media for up to seven days. The results for AlamarBlue assay and PicoGreen assay are shown in Figure 4. The AlamarBlue assay results showed that the measured mean intensities increased from Day 1 to Day 7, which indicated an increased metabolic activity with culture time. There was an exception for A1W4, which showed a drop from Day 4 to Day 7. At Day 7, no significant difference was observed between the test groups. For the PicoGreen assay, a similar trend was observed as the mean value of λ-DNA ascended from Day 1 to Day 7. At Day 7, there was no difference between the test groups. The multiphoton confocal images of the stem cells (F-actin in green and cell nuclei in blue) cultured on the 3D printed scaffolds for seven days are given in Figure 5. F-actin filaments were aligned with the printed struts that formed the scaffolds, and this alignment was more pronounced in the curved regions in wavy scaffolds.

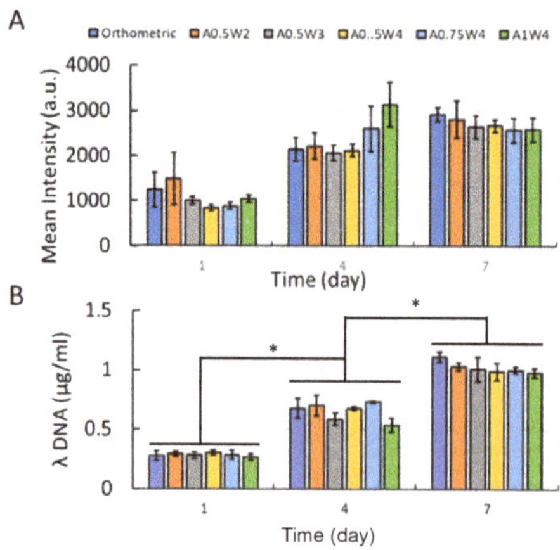

Figure 4. (**A**) AlamarBlue assay results; (**B**) PicoGreen assay results. (* $p < 0.005$).

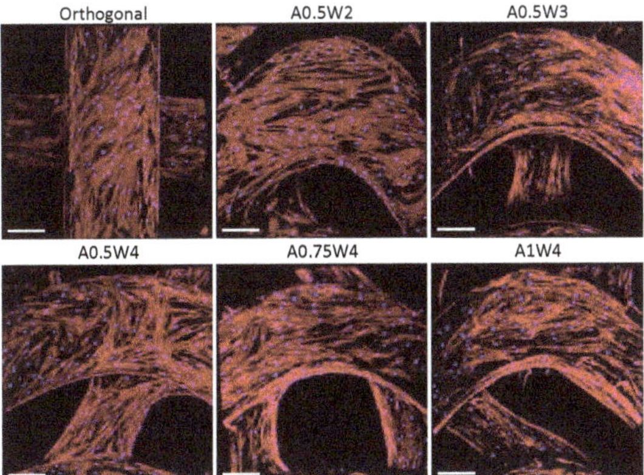

Figure 5. Multiphoton confocal images of the human mesenchymal stem cells (hMSCs) cultured on the scaffolds for seven days. Cells were stained for F-actin (red) and nuclei (blue). Scale bars are 200 microns.

3.4. *Differentiation Study*

Osteogenic differentiation of hMSCs cultured on the 3D printed scaffolds was studied for up to 21 days. Figure 6 shows the results from alizarin red (AR) staining and assay. The scaffolds with wavy designs showed more staining (Figure 6A) and higher values of mean calcium deposition (Figure 6B). The value of the mean calcium deposition in wavy groups was in the range of 2.5 to six times higher than that of the orthogonal group. Specifically, the average calcium deposition was equal to 9.33 ± 0.98 mM for A0.75W4, 8.14 ± 2.86 mM for A0.5W2, 7.60 ± 1.65 mM for A1W4, 6.12 ± 3.07 mM for A0.5W4, 3.96 ± 2.06 mM for A0.5W3, and 1.53 ± 0.10 mM for orthogonal scaffolds, in descending order. ALP activity assay results, at Culture Days 14 and 21, are given in Figure 7. Our results showed an increase in ALP activity for all sample groups from Day 14 to Day 21, and the ALP activity of the wavy scaffolds was higher than that of the orthogonal group at both Day 14 and Day 21 (Figure 7). At Day 14, A0.5W3 (13.16 ± 3.17 a.u.) was significantly higher than the orthogonal group (5.96 ± 1.58 a.u.). At Day 21, A0.5W2 (46.83 ± 7.90 a.u.) and A0.5W3 (45.51 ± 4.20 a.u.) were much higher than that of the orthogonal group (32.31 ± 0.89 a.u.). Representative fluorescent images showing vinculin staining at Day 7 are shown in Figure 8. We observed more pronounced vinculin fibers that were aligned with the wavy struts for wavy scaffolds as compared to diffused and randomly oriented vinculin for the orthogonal scaffold. Figure 9 shows the representative confocal images of the hMSCs cultured on 3D printed scaffolds, in which cells were stained for osteocalcin (OC, green), F-actin (red), and nuclei (blue) at Culture Day 14 and 21. Osteocalcin staining was more pronounced on curved struts as compared to linear struts.

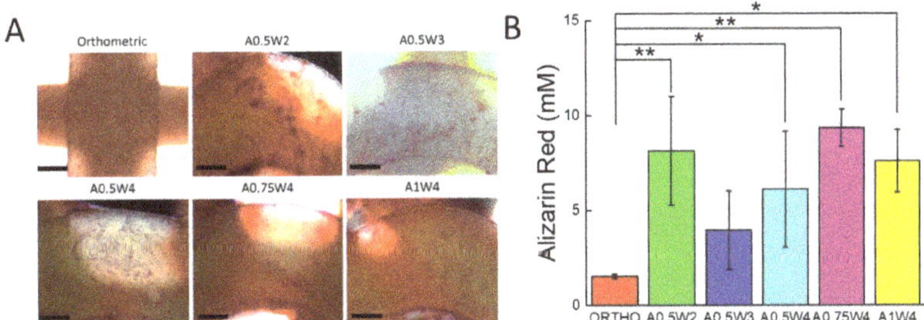

Figure 6. (**A**) Optical microscopy images of the hMSCs stained for alizarin red (red) after culture in osteogenic induction media for 21 days. Scale bars are 200 microns. (**B**) Alizarin red concentration indicating calcium deposition at Day 21. (* $p < 0.15$, ** $p < 0.05$, for $n = 3$).

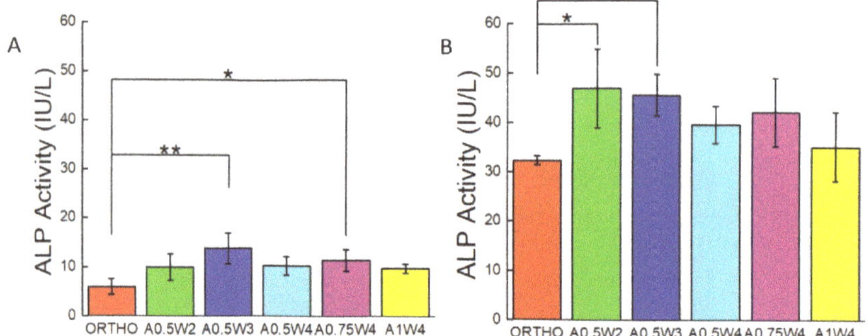

Figure 7. Alkaline Phosphatase (ALP) activity assay results for: (**A**) Day 14 and (**B**) Day 21 (* $p < 0.15$, ** $p < 0.05$, for $n = 3$).

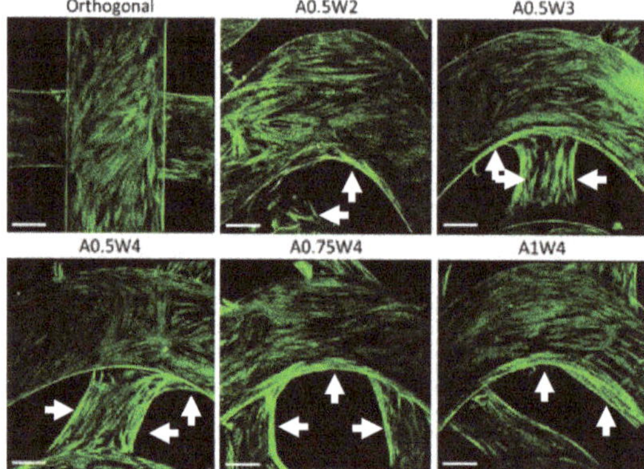

Figure 8. Multiphoton confocal images of hMSCs that are stained for vinculin (green) at Day 7.

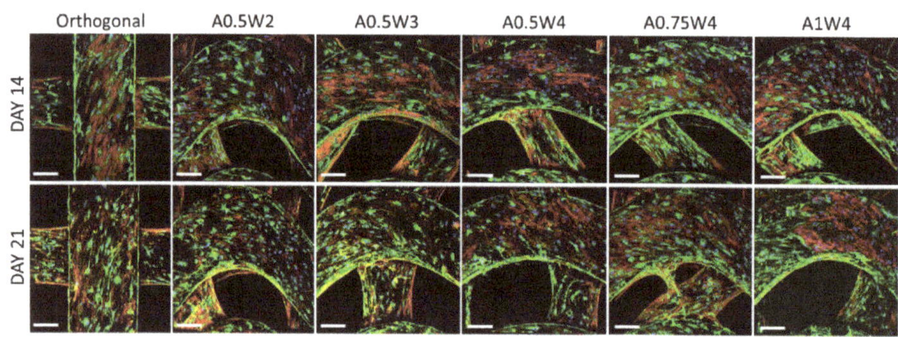

Figure 9. Multiphoton confocal images for hMSCs that were cultured in osteogenic induction media for 14 (**top** row) and 21 days (**bottom** row). Cells were immunostained for osteocalcin (green) and stained for F-actin (red) and cell nuclei (blue). Scale bars are 200 microns.

4. Discussion

In this study, we used extrusion based DIW printing technology to fabricate PCL scaffolds. DIW allowed us to 3D print scaffolds directly from PCL pellets, which were melted within and extruded from a steel syringe attached to the print head. PCL was selected as a model polymer as it is a "Generally

Recognized As Safe" (GRAS) polymer by the U.S. Food and Drug Administration and widely used to 3D print tissue engineering scaffolds for both in vitro and in vivo studies [27,35,75]. We used human adult mesenchymal stem cells (hMSCs) as the main cell line due to their ability to proliferate, migrate, and differentiate into a wide range of tissue specific phenotypes including bone, cartilage, and muscle. Stem cells are known to feel and respond to their microenvironment (matrix stiffness, topography, and bioactivity) by regulating their behavior [56,60]. Here, we focused on the topography, or scaffold architecture. To investigate the effects of 3D scaffold architecture on stem cell osteogenesis, we constructed scaffolds using struts in sinusoidal waveforms, systematically varying the amplitude and the wavelength (Figures 1 and 2, Table 1). The sinusoidal waveform design created highly curved strut surfaces forming 3D scaffolds with wavy patterns. Our motivation to create wavy scaffolds was based on previous studies, which clearly showed the importance of substrate curvature on stem cell osteogenesis [69,70].

The minimum wavelength and amplitude achievable for a strut size around 500 µm were 2 mm and 0.5 mm (A0.5W2). While keeping the amplitude constant at 0.5 mm, the wavelength was increased to 3 mm (A0.5W3) and 4 mm (A0.5W4). For the 4 mm wavelength, the amplitude was increased to 0.75 mm (A0.75W4) and 1 mm (A1W4). These geometrical constraints allowed us to create scaffolds with an average strut-to-strut distance of approximately 350 µm (Figure 2, Table 1). Note that the effect of pore size in bone scaffolds has been well studied [15,24,76–81], and a minimum pore size of ~150 µm is usually required for cell migration and tissue ingrowth [3,82–84]. We then investigated the effect of scaffold design on mechanical properties of the scaffolds (Figure 3). The compressive modulus (E) values were determined by the design, i.e., strut-to-strut contacts between layers, and the overall scaffold porosity. E values were significantly the highest for orthogonal scaffolds (12.5 MPa) mainly because these scaffolds inherently displayed more strut-to-strut contacts, considering that this design had 16 struts per layer, whereas all of the wavy designs had 15 struts per layer. This design also had one of the lowest porosities with ~56%. When wavy scaffolds were compared, A0.5W4 showed the significantly highest porosity (~62%) corresponding to the significantly lowest E value of 9.5 MPa followed by A0.75W4 (58%, 10.7 MPa), A1W4 (57% 11.3MPa), and A0.5W3 (56%, 11.5 MPa). A0.5W2 (56%, 10.5 MPa) was an exception and did not follow the trend. This was due to reduced strut-to-strut contacts due to the design (Figure 2). Although the overall scaffold modulus determines the mechanical support level that a scaffold can provide when implanted, it did not affect the stem cell behavior in our study. This is because the stem cells feel the mechanics of the individual struts (which was uniform for all scaffold groups) that they reside on when seeded on to the scaffolds [74].

First, the growth study was conducted to determine the attachment and proliferation of the hMSCs cultured on our scaffolds. The metabolic activities of the cells were not significantly different from each other at each culture day, but increased significantly with culture day, reaching a maximum at Day 7 (Figure 4A). The same trend was observed when the DNA was quantified (Figure 4B). Note that this trend was not true for the A1W4 and A0.75W4 sample groups, for which the metabolic activity reached a maximum at Day 4 and did not change significantly at Day 7. Yet, the DNA count did not show this unexpected trend for these two sample groups, which represented the cell proliferation more accurately. F-actin staining at Day 7 confirmed that cells attached onto the struts and formed confluent layers at Day 7, taking the shape of the struts. Cells on wavy scaffolds were highly elongated, especially on the curved edges with well-defined F-actin filaments aligned with the scaffold curvature as compared to much bulkier cells on orthogonal scaffolds (Figure 5). In addition, stem cells on wavy scaffolds showed mature vinculin (focal adhesion marker) patches as compared to diffused vinculin staining of cells on orthogonal scaffolds at Day 7 (Figure 8). Focal adhesion is a vital step in osteogenesis [85] in which vinculin directs the interaction between talin and actin to direct the focal adhesion process [86]. We investigated if these significant changes in stem cell morphology, F-actin expression, and focal adhesion on wavy scaffolds as compare to orthogonal scaffolds correlated with stem cell osteogenesis on wavy scaffolds. It was also noted that the curvature had a direct effect on cell proliferation, and studies have shown that curvature induced contractility enhances proliferation and cell growth [87–89]. In our study, we did not observe a significant difference in proliferation between sample groups. This was not contradictory to the literature as each of our wavy scaffolds displayed both concave and convex curvature, and the overall cellular behavior was collective rather than distinct for each type of curvature.

The differentiation studies were conducted after the cells reached a confluent state at Day 7, as shown by the growth studies (Figures 4 and 5). At Day 7, the growth media was replaced with osteogenic induction media, and cells were cultured for 14 additional days in induction media, a total of 21 days in culture. To assess the osteogenic differentiation of hMSCs, we quantified calcium deposition and ALP activity and performed immunostaining for osteocalcin. The AR assay was used to probe the deposition of calcium. Optical microscope images revealed that wavy scaffolds showed more stained regions than the orthogonal group. When quantified, all the wavy scaffolds showed higher calcium deposition than the orthogonal group, and in particular, two groups (A0.5W2 and A0.75W4) showed significantly higher calcium deposition (Figure 6). These results indicated that the overall contribution of the curvature on these two scaffolds on cellular contractility induced calcium deposition was the highest. ALP is a well known biological marker for stem cell osteogenesis [90]. ALP activity increased significantly for all of the scaffold groups from Day 14 to Day 21 (Figure 7). All the wavy groups showed higher ALP activity than the orthogonal group. However, the differences between wavy groups and the orthogonal group were not as significant as the results from the AR assay. This could be because the ALP expressed at earlier stages of the osteogenesis process. At Day 14, A0.5W3 showed significantly higher ALP activity ($p < 0.05$) when compared to the orthogonal group. At Day 21, both A0.5W2 and A0.5W3 were substantially higher than the orthogonal group ($p < 0.15$). To supplement our quantitative differentiation assays, we performed OC immunostaining (Figure 9) as a marker for osteogenesis. Qualitatively, we observed increasing OC staining with culture day, and wavy scaffolds showed more OC staining, in particular in the curved regions of the scaffolds. The enhanced osteogenesis behavior on wavy scaffolds could be explained as the effect of the curvature, which led to a highly aligned and stretched cellular morphology (Figure 5) with mature focal adhesions (Figure 8), leading to highly contractile cells promoting osteogenesis. We strongly believe that our results clearly showed the importance of scaffold architecture on hMSC osteogenesis and would help to develop novel scaffold architectures for bone tissue regeneration.

5. Conclusions

In this study, we developed 3D printed PCL scaffolds with wavy or linear patterns to investigate the effects of a wavy scaffold architecture on the osteogenic differentiation of hMSCs. When cultured in growth media, hMSCs attached and proliferated, forming confluent layers on the scaffolds within seven days. We found that hMSCs spread by taking the shape of the curved surfaces and exhibited elongated F-actin filaments and mature focal adhesion sites (vinculin staining). In contrast, hMSCs were bulkier in shape and showed dispersed vinculin staining on the orthogonal scaffold. We found that hMSCs showed significantly higher calcium deposition, higher ALP activity, and significantly pronounced osteocalcin staining when cultured on wavy scaffolds as compared to orthogonal scaffolds. These results are important in that they clearly showed the importance of scaffold architecture on hMSC osteogenesis and may provide guidance on novel bone scaffold/graft design for pre-clinical and clinical applications.

Author Contributions: S.J. and M.G. designed the experiments, analyzed the data, and wrote the manuscript. S.J. conducted the experiments. All authors have read and agreed to the published version of the manuscript.

Funding: This work is funded by the National Science Foundation Award Number DMR-1714882 (M.G.) and the Faculty Seed Grant from the Center for Engineering MechanoBiology (CEMB), an NSF Science and Technology Center, under Grant Agreement CMMI: 15-48571. Any opinions, findings, and conclusions or recommendations expressed in this material are those of the author(s) and do not necessarily reflect the views of the National Science Foundation.

Acknowledgments: The authors acknowledge Andrew House and Chya-Yan for their help with micro-CT imaging.

Conflicts of Interest: The authors declare no conflict of interest.

References

1. Tang, D.; Tare, R.S.; Yang, L.Y.; Williams, D.F.; Ou, K.L.; Oreffo, R.O. Biofabrication of bone tissue: Approaches, challenges and translation for bone regeneration. *Biomaterials* **2016**, *83*, 363–382. [CrossRef] [PubMed]
2. Oryan, A.; Alidadi, S.; Moshiri, A.; Maffulli, N. Bone regenerative medicine: Classic options, novel strategies, and future directions. *J. Orthop. Surg. Res.* **2014**, *9*, 18. [CrossRef] [PubMed]
3. Bose, S.; Roy, M.; Bandyopadhyay, A. Recent advances in bone tissue engineering scaffolds. *Trends Biotechnol.* **2012**, *30*, 546–554. [CrossRef] [PubMed]
4. Mourino, V.; Boccaccini, A.R. Bone tissue engineering therapeutics: Controlled drug delivery in three-dimensional scaffolds. *J. R. Soc. Interface* **2010**, *7*, 209–227. [CrossRef]
5. Nandi, S.; Roy, S.; Mukherjee, P.; Kundu, B.; De, D.; Basu, D. Orthopaedic applications of bone graft & graft substitutes: A review. *Indian J. Med. Res.* **2010**, *132*, 15–30.
6. Elsalanty, M.E.; Genecov, D.G. Bone grafts in craniofacial surgery. *Craniomaxillofac. Trauma Reconstr.* **2009**, *2*, 125–134. [CrossRef]
7. Albrektsson, T.; Johansson, C. Osteoinduction, osteoconduction and osseointegration. *Eur. Spine J.* **2001**, *10*, S96–S101. [CrossRef]
8. Janicki, P.; Schmidmaier, G. What should be the characteristics of the ideal bone graft substitute? Combining scaffolds with growth factors and/or stem cells. *Injury* **2011**, *42*, S77–S81. [CrossRef]
9. Kucharska, M.; Butruk, B.; Walenko, K.; Brynk, T.; Ciach, T. Fabrication of in-situ foamed chitosan/β-TCP scaffolds for bone tissue engineering application. *Mater. Lett.* **2012**, *85*, 124–127. [CrossRef]
10. Chen, W.; Zhou, H.; Tang, M.; Weir, M.D.; Bao, C.; Xu, H.H.K. Gas-foaming calcium phosphate cement scaffold encapsulating human umbilical cord stem cells. *Tissue Eng. Part A* **2012**, *18*, 816–827. [CrossRef]
11. Costantini, M.; Barbetta, A. 6-Gas foaming technologies for 3D scaffold engineering. In *Functional 3D Tissue Engineering Scaffolds*; Deng, Y., Kuiper, J., Eds.; Woodhead Publishing: Sawston, UK, 2018; pp. 127–149. [CrossRef]
12. Mikos, A.G.; Thorsen, A.J.; Czerwonka, L.A.; Bao, Y.; Langer, R.; Winslow, D.N.; Vacanti, J.P. Preparation and characterization of poly(l-lactic acid) foams. *Polymer* **1994**, *35*, 1068–1077. [CrossRef]
13. Sola, A.; Bertacchini, J.; D'Avella, D.; Anselmi, L.; Maraldi, T.; Marmiroli, S.; Messori, M. Development of solvent-casting particulate leaching (SCPL) polymer scaffolds as improved three-dimensional supports to mimic the bone marrow niche. *Mater. Sci. Eng. C* **2019**, *96*, 153–165. [CrossRef] [PubMed]
14. Thadavirul, N.; Pavasant, P.; Supaphol, P. Development of polycaprolactone porous scaffolds by combining solvent casting, particulate leaching, and polymer leaching techniques for bone tissue engineering. *J. Biomed. Mater. Res. Part A* **2014**, *102*, 3379–3392. [CrossRef]
15. Cao, H.; Kuboyama, N. A biodegradable porous composite scaffold of PGA/β-TCP for bone tissue engineering. *Bone* **2010**, *46*, 386–395. [CrossRef] [PubMed]
16. Liao, C.-J.; Chen, C.-F.; Chen, J.-H.; Chiang, S.-F.; Lin, Y.-J.; Chang, K.-Y. Fabrication of porous biodegradable polymer scaffolds using a solvent merging/particulate leaching method. *J. Biomed. Mater. Res.* **2002**, *59*, 676–681. [CrossRef] [PubMed]
17. Nam, Y.S.; Park, T.G. Porous biodegradable polymeric scaffolds prepared by thermally induced phase separation. *J. Biomed. Mater. Res.* **1999**, *47*, 8–17. [CrossRef]
18. Akbarzadeh, R.; Yousefi, A.M. Effects of processing parameters in thermally induced phase separation technique on porous architecture of scaffolds for bone tissue engineering. *J. Biomed. Mater. Res. B Appl. Biomater.* **2014**, *102*, 1304–1315. [CrossRef]
19. Sultana, N.; Wang, M. Fabrication of HA/PHBV composite scaffolds through the emulsion freezing/freeze-drying process and characterisation of the scaffolds. *J. Mater. Sci. Mater. Med.* **2007**, *19*, 2555. [CrossRef]
20. Whang, K.; Tsai, D.C.; Nam, E.K.; Aitken, M.; Sprague, S.M.; Patel, P.K.; Healy, K.E. Ectopic bone formation via rhBMP-2 delivery from porous bioabsorbable polymer scaffolds. *J. Biomed. Mater. Res.* **1998**, *42*, 491–499. [CrossRef]
21. Prabhakaran, M.P.; Venugopal, J.; Ramakrishna, S. Electrospun nanostructured scaffolds for bone tissue engineering. *Acta Biomater.* **2009**, *5*, 2884–2893. [CrossRef]

22. Yoshimoto, H.; Shin, Y.M.; Terai, H.; Vacanti, J.P. A biodegradable nanofiber scaffold by electrospinning and its potential for bone tissue engineering. *Biomaterials* **2003**, *24*, 2077–2082. [CrossRef]
23. Li, C.; Vepari, C.; Jin, H.-J.; Kim, H.J.; Kaplan, D.L. Electrospun silk-BMP-2 scaffolds for bone tissue engineering. *Biomaterials* **2006**, *27*, 3115–3124. [CrossRef] [PubMed]
24. Bose, S.; Vahabzadeh, S.; Bandyopadhyay, A. Bone tissue engineering using 3D printing. *Mater. Today* **2013**, *16*, 496–504. [CrossRef]
25. Liaw, C.Y.; Guvendiren, M. Current and emerging applications of 3D printing in medicine. *Biofabrication* **2017**, *9*, 024102. [CrossRef]
26. Jakus, A.E.; Rutz, A.L.; Jordan, S.W.; Kannan, A.; Mitchell, S.M.; Yun, C.; Koube, K.D.; Yoo, S.C.; Whiteley, H.E.; Richter, C.-P.; et al. Hyperelastic "bone": A highly versatile, growth factor–free, osteoregenerative, scalable, and surgically friendly biomaterial. *Sci. Transl. Med.* **2016**, *8*, 358ra127. [CrossRef]
27. Kang, H.W.; Lee, S.J.; Ko, I.K.; Kengla, C.; Yoo, J.J.; Atala, A. A 3D bioprinting system to produce human-scale tissue constructs with structural integrity. *Nat. Biotechnol.* **2016**, *34*, 312–319. [CrossRef]
28. Yan, Y.; Chen, H.; Zhang, H.; Guo, C.; Yang, K.; Chen, K.; Cheng, R.; Qian, N.; Sandler, N.; Zhang, Y.S.; et al. Vascularized 3D printed scaffolds for promoting bone regeneration. *Biomaterials* **2019**, *190–191*, 97–110. [CrossRef]
29. Hutmacher, D.W.; Schantz, T.; Zein, I.; Ng, K.W.; Teoh, S.H.; Tan, K.C. Mechanical properties and cell cultural response of polycaprolactone scaffolds designed and fabricated via fused deposition modeling. *J. Biomed. Mater. Res.* **2001**, *55*, 203–216. [CrossRef]
30. Schantz, J.-T.; Brandwood, A.; Hutmacher, D.W.; Khor, H.L.; Bittner, K. Osteogenic differentiation of mesenchymal progenitor cells in computer designed fibrin-polymer-ceramic scaffolds manufactured by fused deposition modeling. *J. Mater. Sci. Mater. Med.* **2005**, *16*, 807–819. [CrossRef]
31. Korpela, J.; Kokkari, A.; Korhonen, H.; Malin, M.; Närhi, T.; Seppälä, J. Biodegradable and bioactive porous scaffold structures prepared using fused deposition modeling. *J. Biomed. Mater. Res. Part B Appl. Biomater.* **2012**, *101B*, 610–619. [CrossRef]
32. Chen, X.; Gao, C.; Jiang, J.; Wu, Y.; Zhu, P.; Chen, G. 3D printed porous PLA/nHA composite scaffolds with enhanced osteogenesis and osteoconductivity in vivo for bone regeneration. *Biomed. Mater.* **2019**, *14*, 065003. [CrossRef] [PubMed]
33. Ji, S.; Dube, K.; Chesterman, J.P.; Fung, S.L.; Liaw, C.Y.; Kohn, J.; Guvendiren, M. Polyester-based ink platform with tunable bioactivity for 3D printing of tissue engineering scaffolds. *Biomater. Sci.* **2019**, *7*, 560–570. [CrossRef] [PubMed]
34. Russias, J.; Saiz, E.; Deville, S.; Gryn, K.; Liu, G.; Nalla, R.K.; Tomsia, A.P. Fabrication and in vitro characterization of three-dimensional organic/inorganic scaffolds by robocasting. *J. Biomed. Mater. Res. Part A* **2007**, *83A*, 434–445. [CrossRef] [PubMed]
35. Williams, J.M.; Adewunmi, A.; Schek, R.M.; Flanagan, C.L.; Krebsbach, P.H.; Feinberg, S.E.; Hollister, S.J.; Das, S. Bone tissue engineering using polycaprolactone scaffolds fabricated via selective laser sintering. *Biomaterials* **2005**, *26*, 4817–4827. [CrossRef] [PubMed]
36. Lan, P.X.; Lee, J.W.; Seol, Y.-J.; Cho, D.-W. Development of 3D PPF/DEF scaffolds using micro-stereolithography and surface modification. *J. Mater. Sci. Mater. Med.* **2009**, *20*, 271–279. [CrossRef] [PubMed]
37. Bian, W.; Li, D.; Lian, Q.; Zhang, W.; Zhu, L.; Li, X.; Jin, Z. Design and fabrication of a novel porous implant with pre-set channels based on ceramic stereolithography for vascular implantation. *Biofabrication* **2011**, *3*, 034103. [CrossRef]
38. Ronca, A.; Ambrosio, L.; Grijpma, D.W. Preparation of designed poly(d,l-lactide)/nanosized hydroxyapatite composite structures by stereolithography. *Acta Biomater.* **2013**, *9*, 5989–5996. [CrossRef]
39. Dean, D.; Jonathan, W.; Siblani, A.; Wang, M.O.; Kim, K.; Mikos, A.G.; Fisher, J.P. Continuous Digital Light Processing (cDLP): Highly accurate additive manufacturing of tissue engineered bone scaffolds. *Virtual Phys. Prototyp.* **2012**, *7*, 13–24. [CrossRef]
40. Dean, D.; Mott, E.; Luo, X.; Busso, M.; Wang, M.O.; Vorwald, C.; Siblani, A.; Fisher, J.P. Multiple initiators and dyes for continuous Digital Light Processing (cDLP) additive manufacture of resorbable bone tissue engineering scaffolds. *Virtual Phys. Prototyp.* **2014**, *9*, 3–9. [CrossRef]
41. Egan, P.F.; Bauer, I.; Shea, K.; Ferguson, S.J. Mechanics of Three-Dimensional Printed Lattices for Biomedical Devices. *J. Mech. Des.* **2019**, *141*, 031703. [CrossRef]

42. Zhang, Y.; Tse, C.; Rouholamin, D.; Smith, P.J. Scaffolds for tissue engineering produced by inkjet printing. *Cent. Eur. J. Eng.* **2012**, *2*, 325–335. [CrossRef]
43. Guvendiren, M.; Molde, J.; Soares, R.M.D.; Kohn, J. Designing Biomaterials for 3D Printing. *ACS Biomater. Sci. Eng.* **2016**, *2*, 1679–1693. [CrossRef] [PubMed]
44. Giannitelli, S.M.; Accoto, D.; Trombetta, M.; Rainer, A. Current trends in the design of scaffolds for computer-aided tissue engineering. *Acta Biomater.* **2014**, *10*, 580–594. [CrossRef] [PubMed]
45. Dias, M.R.; Guedes, J.M.; Flanagan, C.L.; Hollister, S.J.; Fernandes, P.R. Optimization of scaffold design for bone tissue engineering: A computational and experimental study. *Med. Eng. Phys.* **2014**, *36*, 448–457. [CrossRef] [PubMed]
46. Hutmacher, D.W.; Sittinger, M.; Risbud, M.V. Scaffold-based tissue engineering: Rationale for computer-aided design and solid free-form fabrication systems. *Trends Biotechnol.* **2004**, *22*, 354–362. [CrossRef]
47. Wang, M.; Yang, N. Three-dimensional computational model simulating the fracture healing process with both biphasic poroelastic finite element analysis and fuzzy logic control. *Sci. Rep.* **2018**, *8*, 6744. [CrossRef]
48. Poh, P.S.P.; Valainis, D.; Bhattacharya, K.; van Griensven, M.; Dondl, P. Optimization of bone scaffold porosity distributions. *Sci. Rep.* **2019**, *9*, 9170. [CrossRef]
49. Egan, P.F.; Shea, K.A.; Ferguson, S.J. Simulated tissue growth for 3D printed scaffolds. *Biomech. Model. Mechanobiol.* **2018**, *17*, 1481–1495. [CrossRef]
50. Pittenger, M.F.; Mackay, A.M.; Beck, S.C.; Jaiswal, R.K.; Douglas, R.; Mosca, J.D.; Moorman, M.A.; Simonetti, D.W.; Craig, S.; Marshak, D.R. Multilineage potential of adult human mesenchymal stem cells. *Science* **1999**, *284*, 143. [CrossRef]
51. Marolt, D.; Knezevic, M.; Novakovic, G.V. Bone tissue engineering with human stem cells. *Stem Cell Res. Ther.* **2010**, *1*, 10. [CrossRef]
52. Seong, J.M.; Kim, B.-C.; Park, J.-H.; Kwon, I.K.; Mantalaris, A.; Hwang, Y.-S. Stem cells in bone tissue engineering. *Biomed. Mater.* **2010**, *5*, 062001. [CrossRef] [PubMed]
53. Yousefi, A.-M.; James, P.F.; Akbarzadeh, R.; Subramanian, A.; Flavin, C.; Oudadesse, H. Prospect of stem cells in bone tissue engineering: A review. *Stem Cells Int.* **2016**, *2016*, 6180487. [CrossRef] [PubMed]
54. Engler, A.J.; Sen, S.; Sweeney, H.L.; Discher, D.E. Matrix elasticity directs stem cell lineage specification. *Cell* **2006**, *126*, 677–689. [CrossRef] [PubMed]
55. Discher, D.E.; Mooney, D.J.; Zandstra, P.W. Growth factors, matrices, and forces combine and control stem cells. *Science* **2009**, *324*, 1673–1677. [CrossRef] [PubMed]
56. Guvendiren, M.; Burdick, J.A. Stiffening hydrogels to probe short- and long-term cellular responses to dynamic mechanics. *Nat. Commun.* **2012**, *3*, 792. [CrossRef] [PubMed]
57. Discher, D.E.; Janmey, P.; Wang, Y.-L. Tissue cells feel and respond to the stiffness of their substrate. *Science* **2005**, *310*, 1139. [CrossRef] [PubMed]
58. Marklein, R.A.; Burdick, J.A. Controlling stem cell fate with material design. *Adv. Mater.* **2010**, *22*, 175–189. [CrossRef]
59. Burdick, J.A.; Vunjak-Novakovic, G. Engineered microenvironments for controlled stem cell differentiation. *Tissue Eng. Part A* **2009**, *15*, 205–219. [CrossRef]
60. Guvendiren, M.; Burdick, J.A. Engineering synthetic hydrogel microenvironments to instruct stem cells. *Curr. Opin. Biotechnol.* **2013**, *24*, 841–846. [CrossRef]
61. Flemming, R.G.; Murphy, C.J.; Abrams, G.A.; Goodman, S.L.; Nealey, P.F. Effects of synthetic micro- and nano-structured surfaces on cell behavior. *Biomaterials* **1999**, *20*, 573–588. [CrossRef]
62. Chen, W.; Shao, Y.; Li, X.; Zhao, G.; Fu, J. Nanotopographical Surfaces for Stem Cell Fate Control: Engineering Mechanobiology from the Bottom. *Nano Today* **2014**, *9*, 759–784. [CrossRef] [PubMed]
63. Dobbenga, S.; Fratila-Apachitei, L.E.; Zadpoor, A.A. Nanopattern-induced osteogenic differentiation of stem cells—A systematic review. *Acta Biomater.* **2016**, *46*, 3–14. [CrossRef] [PubMed]
64. Guvendiren, M.; Burdick, J.A. Stem Cell Response to Spatially and Temporally Displayed and Reversible Surface Topography. *Adv. Healthc. Mater.* **2013**, *2*, 155–164. [CrossRef] [PubMed]
65. Dalby, M.J.; Gadegaard, N.; Tare, R.; Andar, A.; Riehle, M.O.; Herzyk, P.; Wilkinson, C.D.W.; Oreffo, R.O.C. The control of human mesenchymal cell differentiation using nanoscale symmetry and disorder. *Nat. Mater.* **2007**, *6*, 997. [CrossRef]
66. Oh, S.; Brammer, K.S.; Li, Y.S.J.; Teng, D.; Engler, A.J.; Chien, S.; Jin, S. Stem cell fate dictated solely by altered nanotube dimension. *Proc. Natl. Acad. Sci. USA* **2009**, *106*, 2130. [CrossRef]

67. Faia-Torres, A.B.; Charnley, M.; Goren, T.; Guimond-Lischer, S.; Rottmar, M.; Maniura-Weber, K.; Spencer, N.D.; Reis, R.L.; Textor, M.; Neves, N.M. Osteogenic differentiation of human mesenchymal stem cells in the absence of osteogenic supplements: A surface-roughness gradient study. *Acta Biomater.* **2015**, *28*, 64–75. [CrossRef]
68. McBeath, R.; Pirone, D.M.; Nelson, C.M.; Bhadriraju, K.; Chen, C.S. Cell Shape, Cytoskeletal Tension, and RhoA Regulate Stem Cell Lineage Commitment. *Dev. Cell* **2004**, *6*, 483–495. [CrossRef]
69. Ruiz, S.A.; Chen, C.S. Emergence of patterned stem cell differentiation within multicellular structures. *Stem Cells* **2008**, *26*, 2921–2927. [CrossRef]
70. Kilian, K.A.; Bugarija, B.; Lahn, B.T.; Mrksich, M. Geometric cues for directing the differentiation of mesenchymal stem cells. *Proc. Natl. Acad. Sci. USA* **2010**, *107*, 4872–4877. [CrossRef]
71. Guvendiren, M.; Burdick, J.A. The control of stem cell morphology and differentiation by hydrogel surface wrinkles. *Biomaterials* **2010**, *31*, 6511–6518. [CrossRef]
72. Viswanathan, P.; Ondeck, M.G.; Chirasatitsin, S.; Ngamkham, K.; Reilly, G.C.; Engler, A.J.; Battaglia, G. 3D surface topology guides stem cell adhesion and differentiation. *Biomaterials* **2015**, *52*, 140–147. [CrossRef] [PubMed]
73. Kumar, G.; Tison, C.K.; Chatterjee, K.; Pine, P.S.; McDaniel, J.H.; Salit, M.L.; Young, M.F.; Simon, C.G., Jr. The determination of stem cell fate by 3D scaffold structures through the control of cell shape. *Biomaterials* **2011**, *32*, 9188–9196. [CrossRef] [PubMed]
74. Guvendiren, M.; Fung, S.; Kohn, J.; De Maria, C.; Montemurro, F.; Vozzi, G. The control of stem cell morphology and differentiation using three-dimensional printed scaffold architecture. *MRS Commun.* **2017**, *7*, 383–390. [CrossRef] [PubMed]
75. Morrison, R.J.; Hollister, S.J.; Niedner, M.F.; Mahani, M.G.; Park, A.H.; Mehta, D.K.; Ohye, R.G.; Green, G.E. Mitigation of tracheobronchomalacia with 3D-printed personalized medical devices in pediatric patients. *Sci. Transl. Med.* **2015**, *7*, 285ra264. [CrossRef] [PubMed]
76. Xue, W.; Krishna, B.V.; Bandyopadhyay, A.; Bose, S. Processing and biocompatibility evaluation of laser processed porous titanium. *Acta Biomater.* **2007**, *3*, 1007–1018. [CrossRef] [PubMed]
77. Balla, V.K.; Bodhak, S.; Bose, S.; Bandyopadhyay, A. Porous tantalum structures for bone implants: Fabrication, mechanical and in vitro biological properties. *Acta Biomater.* **2010**, *6*, 3349–3359. [CrossRef]
78. Bittner, S.M.; Smith, B.T.; Diaz-Gomez, L.; Hudgins, C.D.; Melchiorri, A.J.; Scott, D.W.; Fisher, J.P.; Mikos, A.G. Fabrication and mechanical characterization of 3D printed vertical uniform and gradient scaffolds for bone and osteochondral tissue engineering. *Acta Biomater.* **2019**, *90*, 37–48. [CrossRef]
79. Fielding, G.A.; Bandyopadhyay, A.; Bose, S. Effects of silica and zinc oxide doping on mechanical and biological properties of 3D printed tricalcium phosphate tissue engineering scaffolds. *Dent. Mater.* **2012**, *28*, 113–122. [CrossRef]
80. Murphy, C.M.; Haugh, M.G.; O'Brien, F.J. The effect of mean pore size on cell attachment, proliferation and migration in collagen-glycosaminoglycan scaffolds for bone tissue engineering. *Biomaterials* **2010**, *31*, 461–466. [CrossRef]
81. Sicchieri, L.G.; Crippa, G.E.; de Oliveira, P.T.; Beloti, M.M.; Rosa, A.L. Pore size regulates cell and tissue interactions with PLGA-CaP scaffolds used for bone engineering. *J. Tissue Eng. Regen. Med.* **2012**, *6*, 155–162. [CrossRef]
82. Karageorgiou, V.; Kaplan, D. Porosity of 3D biomaterial scaffolds and osteogenesis. *Biomaterials* **2005**, *26*, 5474–5491. [CrossRef] [PubMed]
83. Jones, A.C.; Arns, C.H.; Sheppard, A.P.; Hutmacher, D.W.; Milthorpe, B.K.; Knackstedt, M.A. Assessment of bone ingrowth into porous biomaterials using MICRO-CT. *Biomaterials* **2007**, *28*, 2491–2504. [CrossRef] [PubMed]
84. Hollister, S.J. Porous scaffold design for tissue engineering. *Nat. Mater.* **2005**, *4*, 518. [CrossRef] [PubMed]
85. Biggs, M.J.P.; Dalby, M.J. Focal adhesions in osteoneogenesis. *Proc. Inst. Mech. Eng. Part H J. Eng. Med.* **2010**, *224*, 1441–1453. [CrossRef]
86. Humphries, J.D.; Wang, P.; Streuli, C.; Geiger, B.; Humphries, M.J.; Ballestrem, C. Vinculin controls focal adhesion formation by direct interactions with talin and actin. *J. Cell Biol.* **2007**, *179*, 1043–1057. [CrossRef]
87. Bidan, C.M.; Kommareddy, K.P.; Rumpler, M.; Kollmannsberger, P.; Bréchet, Y.J.M.; Fratzl, P.; Dunlop, J.W.C. How linear tension converts to curvature: Geometric control of bone tissue growth. *PLoS ONE* **2012**, *7*, e36336. [CrossRef]

88. Nelson, C.M.; Jean, R.P.; Tan, J.L.; Liu, W.F.; Sniadecki, N.J.; Spector, A.A.; Chen, C.S. Emergent patterns of growth controlled by multicellular form and mechanics. *Proc. Natl. Acad. Sci. USA* **2005**, *102*, 11594. [CrossRef]
89. Rumpler, M.; Woesz, A.; Dunlop, J.W.C.; van Dongen, J.T.; Fratzl, P. The effect of geometry on three-dimensional tissue growth. *J. R. Soc. Interface* **2008**, *5*, 1173–1180. [CrossRef]
90. Golub, E.E.; Boesze-Battaglia, K. The role of alkaline phosphatase in mineralization. *Curr. Opin. Orthop.* **2007**, *18*, 444–448. [CrossRef]

Article

Scaffold-Free Bioprinter Utilizing Layer-By-Layer Printing of Cellular Spheroids

Wesley LaBarge [1], Andrés Morales [2], Daniëlle Pretorius [1], Asher M. Kahn-Krell [1], Ramaswamy Kannappan [1] and Jianyi Zhang [1,*]

1 Department of Biomedical Engineering, School of Medicine, School of Engineering, University of Alabama at Birmingham, Birmingham, AL 35294, USA

2 Department of Mechanical Engineering, School of Engineering, University of Alabama at Birmingham, Birmingham, AL 35294, USA

* Correspondence: jayzhang@uab.edu; Tel.: +1-205-934-8423

Received: 13 June 2019; Accepted: 27 August 2019; Published: 29 August 2019

Abstract: Free from the limitations posed by exogenous scaffolds or extracellular matrix-based materials, scaffold-free engineered tissues have immense clinical potential. Biomaterials may produce adverse responses, interfere with cell–cell interaction, or affect the extracellular matrix integrity of cells. The scaffold-free Kenzan method can generate complex tissues using spheroids on an array of needles but could be inefficient in terms of time, as it moves and places only a single spheroid at a time. We aimed to design and construct a novel scaffold-free bioprinter that can print an entire layer of spheroids at once, effectively reducing the printing time. The bioprinter was designed using computer-aided design software and constructed from machined, 3D printed, and commercially available parts. The printing efficiency and the operating precision were examined using Zirconia and alginate beads, which mimic spheroids. In less than a minute, the printer could efficiently pick and transfer the beads to the printing surface and assemble them onto the 4 × 4 needles. The average overlap coefficient between layers was measured and found to be 0.997. As a proof of concept using human induced pluripotent stem cell-derived spheroids, we confirmed the ability of the bioprinter to place cellular spheroids onto the needles efficiently to print an entire layer of tissue. This novel layer-by-layer, scaffold-free bioprinter is efficient and precise in operation and can be easily scaled to print large tissues.

Keywords: tissue engineering; cell; bioprinting; spheroids

---

1. Introduction

With the development of cutting-edge strategies and techniques for generating complex, functional tissues, bioprinting, or three-dimensional (3D) printing of biological materials, has become a critical tool in tissue engineering [1–7]. Bioprinted and fabricated tissues have immense clinical potential including organ and tissue regeneration, drug testing, organ models for surgical practices, and organ replacement. The conventional method of bioprinting is to print cells with or into a scaffold material. Scaffold printing, with either a synthetic or natural polymer, can provide the necessary structure and microenvironment for proper cell growth and survival [8–10]. This method can be used to dictate how structures are arranged, where cells are positioned in this scaffold, and what sort of factors are present to optimize the growth and function of the cells. Polymers can also be designed to degrade depending on a specific set of environmental conditions or with the introduction of a particular solution. However, with synthetic polymers, biocompatibility could pose problems if their use is not appropriately addressed in in vivo studies, leading to an increased use of natural polymers and hydrogels for bioprinting [11–15]. These types of materials possess the same ability to provide an adequate structure for the cells and can be designed to mimic the in vivo environment better, decreasing the chances of rejection if used in the body.

Cells, however, possess the innate ability to produce their scaffold material in the form of extracellular matrix (ECM) as they grow and interact with one another [16–19]. Determining how to harness this ability and become completely scaffold-free effectively has led to the use of 3D spheroids as building blocks for larger tissues [20]. Spheroids can easily fuse together when in close proximity to or touching another spheroid. The efficacy of spheroids as building blocks for larger tissues has been demonstrated using scaffold materials to provide a structure or a particular shape for the proper fusing of spheroids [21]. However, this method still relies on additional materials for tissue generation. This led to the invention of the Kenzan bioprinting method.

The Kenzan method utilizes an array of stainless steel microneedles to provide a structure that the spheroids can be punctured onto to facilitate the fusion process without additional materials [22]. This method allows an actual scaffold-free tissue to be generated from various cell types for a custom-designed tissue. Regenova (Cyfuse Biomedical, Tokyo, Japan) is the only bioprinter on the tissue engineering market which utilizes the Kenzan method. Despite its young age and high cost, it has already been used to generate functional tissues from a multitude of cell types and lineages [23–26]. This device, however, places one spheroid at a time on the needles, requiring a longer printing process as the size of the tissue increases. Having the ability to build larger, more clinically relevant tissues in a shorter length of time using this method would be very beneficial for various fields of medicine and clinical research.

We hypothesize that a device which can print spheroids with a layer-by-layer method will not only reduce the manufacturing time of clinically relevant-sized tissues but also assists in advancing the field of tissue engineering by introducing an affordable alternative for printing scaffold-free, spheroid tissues. The research described here is focused on the design, construction, and testing of such a bioprinter. We also show that we were able to build a working and affordable prototype which efficiently and accurately transfers cellular spheroids to a needle array, one 3 mm × 3 mm × 1 mm layer at a time.

2. Materials and Methods

2.1. Preparation of Needle Arrays

The printing method used by the utilized bioprinter relies on the accurate and precise placement of needles for the proper fusion of spheroids into a single tissue. Needle arrays (Figure 1Aii,Fi) were made using sterile stainless steel needles that were 180 μm in diameter and ~15 mm in length. To aid in the precise placement of the needles, a custom stainless steel plate (Figure 1C,Fii) that possessed an array of through holes, machined using a laser drilling technique, was used. With this plate, needles were placed into a 4 × 4 array, each with a pitch of 800 μm. Once the needles were loaded into the plate, the exposed blunt ends were placed in a silicone solution (Sylgard 184, Dow Corning, Midland, MI, USA) and cured at 70 °C for 2 h to hold the needles in place during the printing process effectively. After curing, the needle array assembly was autoclaved and then kept until ready for use with the bioprinter.

2.2. Bioprinter Design and Construction

The bioprinter (Figure 1A) was designed to fit easily into a biosafety cabinet and operate under sterile conditions. It is 10 inches tall and 8 inches wide, with a depth of 13 inches. The bioprinter was entirely modeled in Fusion 360 (Autodesk) before manufacturing the required pieces for the device. The model of the bioprinter as well as the built bioprinter are shown in Figure 1. The bioprinter consists of three main parts: a spheroid print head (Figure 1D), a spheroid bath stage (Figure 1E), and a needle array bath (Figure 1F). The spheroid print head (Figure 1Ai,Di) was designed to hold a 4 × 4 array of spheroids in a single layer. Each spheroid is approximately 800 μm apart when held on the end of the print head. The print head was machined from 316 stainless steel, and through holes were cut into the bottom of the print head with a diameter of approximately 550 μm, allowing the needles to pass through but preventing spheroid aspiration (Figure 1B). The beads were picked up by a vacuum pump (12V vacuum pump, SparkFun, Niwot, CO, USA) connected to the print head. The spheroid bath stage functioned to hold the reservoir/container of beads during the printing process. The reservoir consisted of a watch glass made from polytetrafluoroethylene (PTFE), whose concave bottom allowed for the beads to collect in the center after each layer was picked up. The beads were printed onto the needles which were held inside the needle array bath. Each piece of the printer can either be sterilized by autoclave or ethylene oxide.

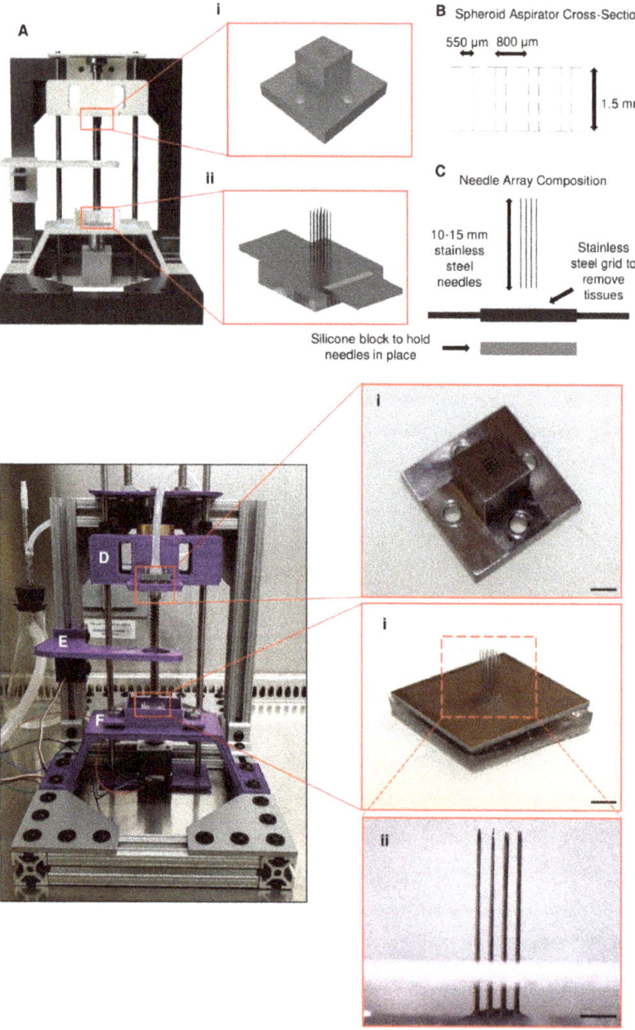

Figure 1. Design and construction of the bioprinter and its components. (**A**) A model of the bioprinter was first made using the CAD modeling software. The crucial components of the bioprinter are (i) the layer-by-layer print head and (ii) the needle array. (**B**) The print head has approximately 550 μm through holes cut into the bottom in a 4 × 4 arrangement, with a space of 800 μm between each hole. (**C**) The needle array is made up of 180 μm-diameter stainless steel needles placed 800 μm apart in a matching 4 × 4 arrangement. (**D–F**) The other components were 3D printed and are as follows: (**D**) (i) spheroid print head, (**E**) spheroid bath stage, and (**F**) needle array bath with (i,ii) needle array. Scale bars: (**D**)(i) and (**F**)(i) 5 mm; (**F**)(ii) 2 mm.

2.3. Bioprinting Process

This scaffold-free bioprinter was controlled using an Arduino (Arduino Uno Rev3, Arduino, Somerville, MA, USA) and a custom-written script. It controlled the stepper motor (NEMA 14, Lin Engineering, Morgan Hill, CA, USA) for moving the print head, the servomotor (5V servomotor, Arduino, Somerville, MA, USA) for moving the spheroid stage, the vacuum pump for aspirating the spheroids, and the end stop (Mechanical End Stop, SparkFun, Niwot, CO, USA) mechanism for calibrating the stepper motor prior to beginning the printing process. The bioprinter was designed to first automatically calibrate to ensure that the print head is in the proper position before beginning. Once it has calibrated, the Arduino program prompts the user to load the spheroids into the spheroid stage in the round bottom plate (Figure 2B). Next, the user is asked to input the exact number of layers to print, between one and five. Once this value is entered, the printing process begins (Video S1). First, the print head is lowered down into the spheroid container, and the vacuum pump is activated to aspirate one layer of spheroids (Figure 2C,D). While the vacuum is still on, the print head is raised out of the container, and the spheroid stage is moved out of the way. The print head with spheroids is lowered straight down onto the needles in the array bath (Figure 2E). Once the print head is placed in the correct position, the vacuum pump is turned off, and the spheroids are left on the needles. This process is repeated until all layers have been placed on the needles, with each subsequent layer being placed ~800 µm higher than the previous layer. After all layers are completed, the array bath is removed from the system and placed in the incubator for at least 24 h for fusion to take place. Once the spheroids have adequately fused, the stainless steel plate is used to assist in the removal of the fused tissue for further processing (Figure 2F).

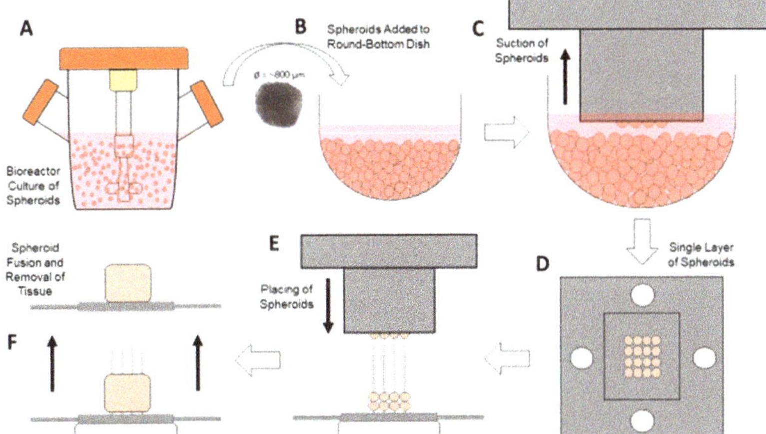

Figure 2. Bioprinting process. Spheroid culture can take place in (**A**) a bioreactor or a standard round-bottom culture plate, and once the spheroids are approximately 800 µm in diameter, they can be placed in (**B**) the round-bottom dish on the spheroid stage. Once the desired number of spheroids are placed in the dish, (**C**) the print head is lowered down to the center of the dish, and the vacuum is turned on, capturing (**D**) a single layer of spheroids. With the vacuum still on, the print head is lowered down to the needles (**E**), placing the spheroids on them. After the required number of layers is achieved, (**F**) the spheroids are allowed to fuse and then later removed for further processing.

2.4. Vacuum and Print Head Testing

Glass beads were used to determine the efficiency of the vacuum and print head system. Glass beads approximately 700 µm in diameter (±10% diameter, Zirconia Beads, BioSpec Products, Bartlesville, OK, USA) were used in place of cultured spheroids because of the number of tests that needed to be completed and the non-sterile conditions of the testing area. This diameter was chosen because it is similar to the 800 µm diameter of the spheroids that our lab has previously produced [27]. These beads were held on the spheroid stage in the PTFE container. To help counteract aggregation during the testing process, the beads were placed in a solution of phosphate-buffered saline with 0.1% Tween 20 (PBST), which also allowed to simulate spheroids in media. To determine the different efficiencies of these experiments, the print head was lowered onto the spheroid stage and the vacuum was turned on, which allowed the print head to pick up the beads. After a given time, the number of aspirated beads were counted and calculated as a percentage of the total number of holes available to collect the beads (e.g., 15 beads aspirated into an array of 16 possible holes equates to an efficiency of 93.8%).

2.5. Preparation of Alginate Beads

Alginate beads were chosen as a reasonable spheroid surrogate to test the ability of the bioprinter to place spheroids on the needle array effectively. As mentioned previously, spheroids that would be used with this system are approximately 800 µm in diameter, so the alginate beads needed to have a similar diameter. These beads were generated by adding a solution of 1.5% sodium alginate (Cat. No. 218295, MP Biomedicals, Irvine, CA, USA) in deionized water (DI) water to a 100 mM solution of calcium chloride (Cat. No. 4901, Sigma Aldrich, St. Louis, MO, USA) in DI water [28]. The alginate solution was added using a 22-gauge serological needle dropwise into the calcium chloride solution which was being stirred using a magnetic stir bar. The beads were separated according to their approximate diameter and measured using image analysis.

2.6. Preparation of Human Induced Pluripotent Stem Cell (hiPSC) Spheroids

Human induced pluripotent stem cells (hiPSCs) were cultured on Matrigel-coated dishes at 37 °C under 5% CO_2 with mTeSR (Stem Cell Technologies, Vancouver, BC, Canada). Fresh medium was added daily until the cells reached 90% confluency. The cells were seeded and grown in suspension according to the manufacturer's protocol [29]. Briefly, the cells were passaged by treating with Gentle Cell Dissociation Reagent (Stem Cell Technologies, Vancouver, BC, Canada) for 7 min. The cells were carefully scraped off the dish and seeded at a density of 5×10^5 cells per mL in 30 mL of mTeSR 3D Seed Media (Stem Cell Technologies, Vancouver, BC, Canada). This solution was then cultured in a 125 mL shaker flask at 70 rpm with daily media additions of 3.4 mL. Prior to their use with the bioprinter, the spheroids were filtered through a 500 µm reversible strainer (pluriSelect, El Cajon, CA, USA) and collected in 5 mL of mTeSR.

2.7. Statistical and Image Analysis

Data are shown in the form mean ± standard error of the mean (SEM). Significance was chosen as $p < 0.05$. This was determined using both one- and two-way analysis of variance (ANOVA) along with the Student's t-test. These analyses were performed utilizing Microsoft Excel's data analysis software package.

For image analysis, ImageJ (Version 1.52o with the addition of plug-ins for colocalization (Colocalization Finder, Version 1.2, Institut de Biologie Moleculaire des Plantes, Strasbourg, France) was used to estimate the overlap coefficient between each layer of beads and to generate representative pictures. Also, it was used in the diameter measurements of the alginate beads that were synthesized. Images and videos in this experiment were taken with a stereomicroscope (Olympus SZ61, Olympus, Center Valley, PA, USA), a phase-contrast microscope (AMG EVOS FL Imaging System, Thermo Fisher Scientific, Waltham, MA, USA), and a Canon camera (Canon EOS Rebel T7i, Canon, Tokyo, Japan).

3. Results

3.1. System Testing for Efficiency of Sphere Capture and Transfer

To test the ability of the vacuum pump system and print head to pick up spheroids, glass beads similar in diameter to the average spheroids that would be used with this printing setup were used.

These beads, according to the supplier's specifications, had a diameter of 700 ± 70 μm on average (Figure 3A), which is very similar to the ~800 μm diameter of the spheroids that would be used. The efficiency of the system was determined by using two separate containers (Beads Only and Beads + PBST) of spheres which were picked up by turning on the vacuum pump system for 10 s at a time, removing the print head (Figure 3B) from the container, and then counting how many spheres had been picked up. After collecting the beads (Figure 3C) various times (n = 28 for Beads Only; n = 18 for Beads + PBST), the efficiencies of sphere capture were determined to be 98.3 ± 0.5% for Beads Only and 98.4 ± 0.6% for Beads + PBST (Figure 3D).

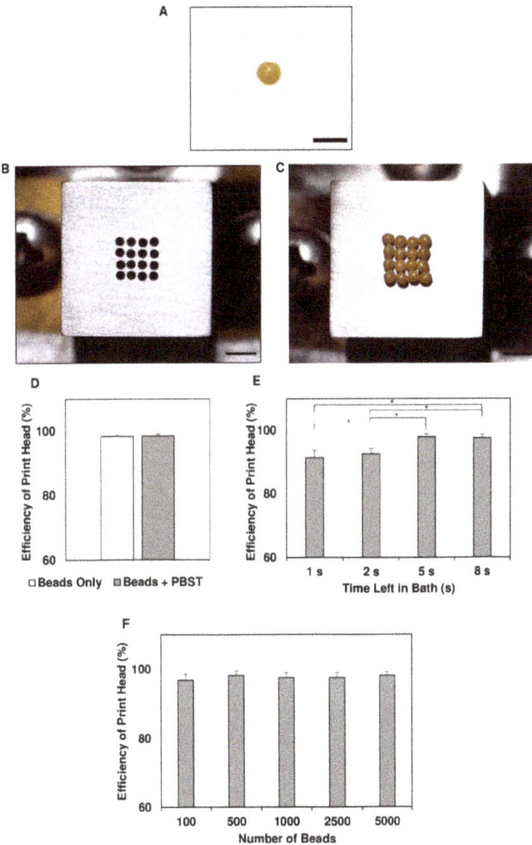

Figure 3. Vacuum and print head capture efficiencies. A substitute was chosen for spheroids because of the non-sterile conditions and multiple tests that needed to be performed. (**A**) Zirconia beads with a diameter (700 ± 70 μm) equivalent to that of the spheroids were used. (**B**,**C**) The print head (B, seen from bottom) was used to pick up (C) a 4 × 4 array of beads in a dry setup. After the print head was placed in the desired container of beads, the vacuum was turned on for 10 s. At the end, the number of beads captured was counted, and (**D**) the efficiencies were calculated. (**E**) Then, the optimal time for the vacuum to be on at which most beads would be captured was determined to be 5 s (* $p < 0.05$, n = 13 for each); (**F**) how the number of beads in the container affected the precision of the bioprinter was also determined. Scale bars: (**A**) 1 mm, (**B**), and (**C**) 2 mm.

To further optimize the printing process, the time during which the print head remained in the bead container was varied (1 s, 2 s, 5 s, and 8 s) to determine the optimal time needed to pick up most of the beads. For each time, the corresponding efficiencies (n = 13 for each time point) were calculated

(Figure 3E). From this test, it was determined that 5 s was the shortest and most efficient time, with an efficiency of 97.8 ± 0.9% which was significantly different ($p < 0.05$) from those measured for times of 1 s (91.4 ± 2.3%) and 2 s (92.6 ± 1.7%) but did not change significantly when the time increased to 8 s (97.5 ± 1.1%) or 10 s.

In addition, due to the fact that as spheroids are removed, more empty space in the spheroid container would be present, a test was done to determine if the number of beads within the container affected the capture efficiency. Beads were added to the dish at amounts of 100, 500, 1000, 2500, and 5000, estimated on a mass basis. The print head was used to capture beads a number of times (n = 10 for each group), and the subsequent efficiencies were determined (average efficiency: 97.6 ± 0.2%) (Figure 3F). There was no significant difference between any amounts of beads tested, showing that as the number of beads decreases in the container, the efficiency does not change.

The repeatability of the bioprinter ability to place the beads in the same area with each subsequent layer was tested to ensure that the spheroids could accurately be placed on the needles when generating tissues. This was determined by placing a full layer of beads into a pre-determined area on a layer of silicone multiple times. After every single layer was placed, an image was taken of the beads, with each placed layer given a different color (Figure 4A). A composite image of each layer aligned together was generated to demonstrate that the bioprinter could place the spheroids in the same area, made apparent by the overlap in colors (see yellow color) (Figure 4B). To estimate the degree of overlap between the layers, a colocalization tool in conjunction with ImageJ was used to calculate the overlap coefficient for each combination of any two layers (Figure 4C). The overlap coefficients between any two layers were greater than 0.99, showing that each layer closely overlapped with the others.

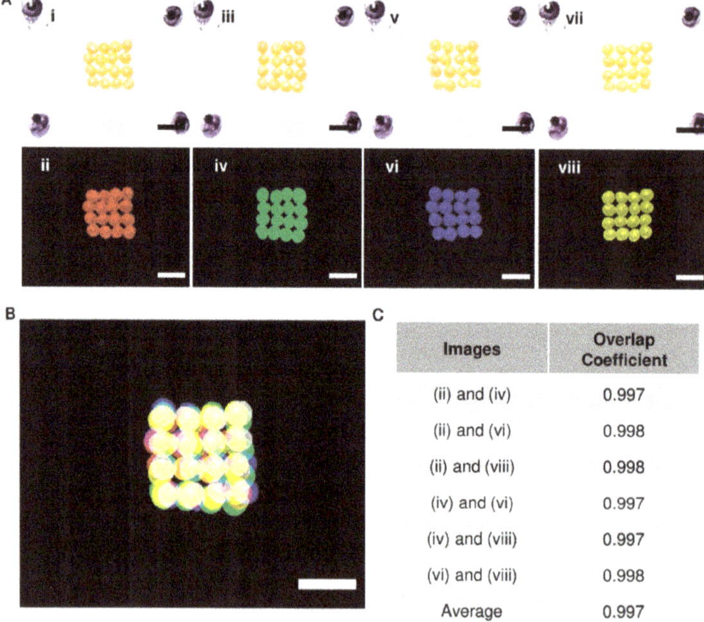

Images	Overlap Coefficient
(ii) and (iv)	0.997
(ii) and (vi)	0.998
(ii) and (viii)	0.998
(iv) and (vi)	0.997
(iv) and (viii)	0.997
(vi) and (viii)	0.998
Average	0.997

Figure 4. Bead layer-by-layer overlap coefficient determination. To determine the precision of the bioprinter, (**A**) layers of beads were added to a predefined area (upper panels), and each run was assigned a color (lower panels). (**B**) Images from individual runs were overlapped using ImageJ. The yellow color indicates the overlap of the beads. (**C**) The overlap coefficient was determined (Colocalization Finder for ImageJ) using the images, and the average overlap coefficient was found to be 0.997. Scale bars: 2 mm.

3.2. *Alginate Bead Formation and Print Head Testing*

Small beads made from alginate were chosen as a surrogate for cellular spheroids for testing the bioprinter's ability to print onto the needle arrays correctly. Alginate beads were synthesized using a dropwise technique (Video S2) and separated according to their diameter. Once these beads were separated, they were collected into a single dish and measured using phase-contrast microscopy (Figure 5A). The average diameter of the beads (n = 27) was determined to be 851 ± 18 μm, which is similar to the desired diameter of 800 μm. The beads were then picked up with the print head (Figure 5B) and transferred to the needles (Figure 5C) using the bioprinting process (Video S3). Placing a single layer of beads took approximately 45 s.

Additionally, to determine how effectively the bioprinter could print multiple layers of spheroids, two layers were printed onto the needles (Figure 5D; Video S4). From these data, it was seen that the bioprinter could successfully print the alginate beads in multiple layers. Our results clearly show that multilayer printing of spheroids is possible using our newly designed bioprinter.

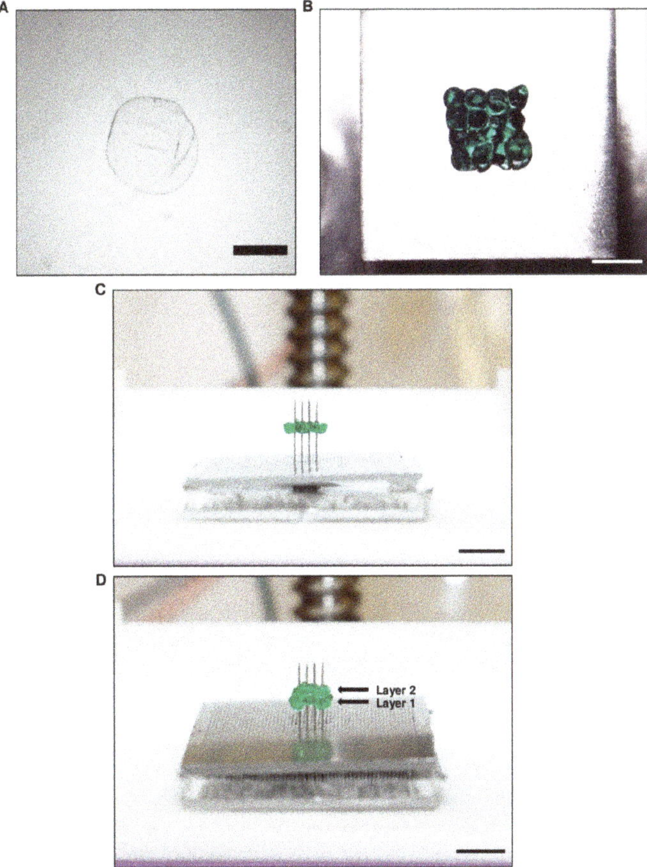

Figure 5. Bioprinter testing with alginate beads. Soft alginate beads were used as a spheroid surrogate as they could easily be punctured and placed on the bioprinter needles. (**A**) The beads were synthesized at an average diameter of 851 ± 15 μm, which was similar to the spheroid diameter. (**B–D**) Picture (**B**) showing beads aspirated onto the end of a 4 × 4 arrangement print head and (**C**) printed onto the needles in a single layer. (**D**) Two layers were also tested to ensure that multiple layers could be printed. Scale bars: (**A**) 500 μm; (B) 2 mm; (**C**) and (**D**) 5 mm.

3.3. Proof-Of-Concept Testing with hiPSC Spheroids

We confirmed the potential of our bioprinter to place cellular spheroids onto the needles in a layer-by-layer fashion. For this, hiPSC spheroids were grown to a similar diameter of, on average, 718 ± 77 µm (Figure 6A, Image S1) and added to the round-bottom container. The bioprinter was started, and a single layer of spheroids was picked up and placed onto the needles. The top (Figure 6B,C) and side (Figure 6D,E) view images show a single layer of spheroids on the needles. These data clearly show that the bioprinter was effectively able to aspirate and transfer a single layer of cellular spheroids onto the needles.

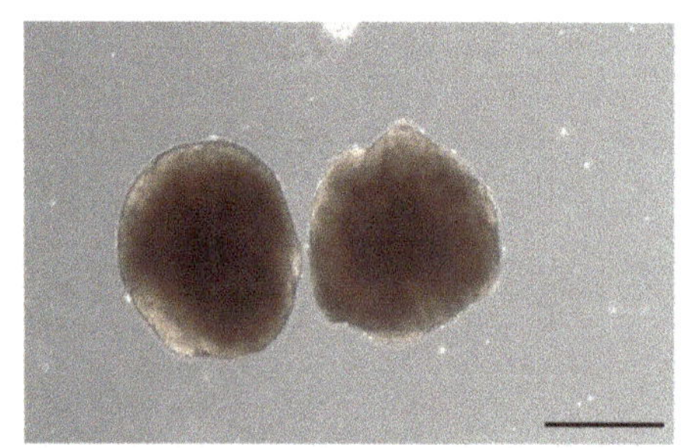

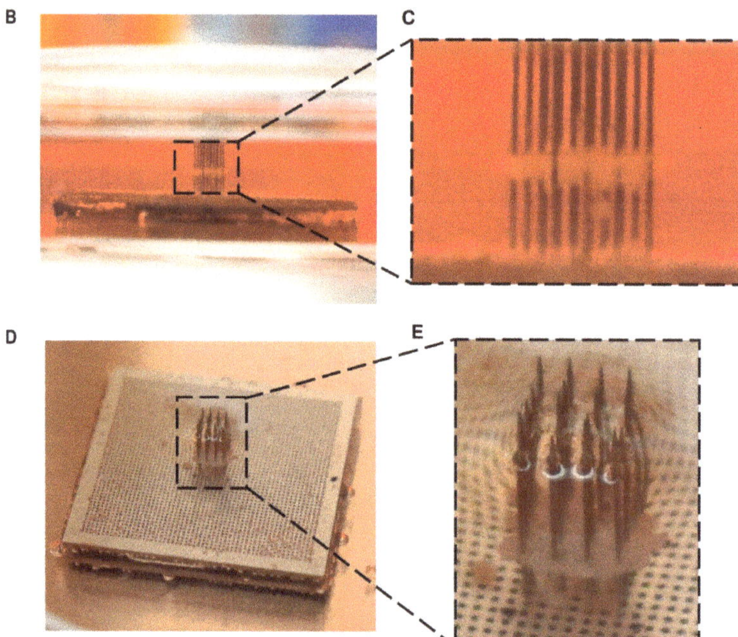

Figure 6. Proof-of-concept testing with human induced pluripotent stem cells (hiPSC) spheroids. hiPSC spheroids were cultured using a rotating culture flask method. (**A**) The spheroids were grown to an average diameter of 718 ± 77 µm. Once enough spheroids were available to be used with the bioprinter to create a single layer, they were loaded into the bioprinter and placed onto the needles. (**B**,**C**) Side view of the hiPSC spheroid layer after placement onto the needles. (**D**,**E**) Top view of the hiPSC spheroid layer. The ability to print cellular spheroids onto the needle array confirms the efficiency of the layer-by-layer printing technique presented here.

4. Discussion

In this study, we were able to construct a novel and affordable bioprinting device with the ability to produce scaffold-free tissues through the utilization of layer-by-layer printing of cellular spheroids. This device could allow the enhanced production of viable tissues to be used in drug testing, translational research, and various clinics.

With the design and development of the layer-by-layer print head, efficiently picking up the spheroids from a reservoir was paramount for the overall success of the bioprinter. Through testing with beads of size similar to that of the spheroids, we were able to achieve nearly 98% efficiency when picking up the beads in a single 4×4 layer. Although the print head designed in this project could only transfer a maximum of 16 beads or spheroids, changes can easily be made to the number and arrangement of holes in the bottom of the print head. This would allow for the construction of both larger tissues as well as custom arrangements to be used in different applications where 3D models of the native tissue microenvironment are needed [30,31]

In addition, we were able to show the reliability of the bioprinter to print in a localized position repeatedly. The lowest overlap coefficient calculated between any two layers was 0.997. The closer to 1.0 this value is, the more colocalized the beads of different layers are. This corresponded to the precise placement of the spheroids on the needles when testing with hiPSC spheroids. This precision will allow for the proper alignment of layers in the construction of native tissue microenvironments as well as any cellular gradients found within tissues. The reproducibility and speed of printing viable tissues are critical for the success of bioprinting in the clinical environment [32,33]. The only commercially available spheroid printing device, Regenova by Cyfuse Biomedical, was used to generate a tubular, vessel-like structure that was made from approximately 500 spheroids [26]. Itoh et al. noted that to place so many spheroids onto the needles, it took the bioprinter approximately 1.3 h (78 min). This corresponds to approximately six spheroids printed per minute. The bioprinter we present here can print 16 spheroids per minute, that is, an over 250% increase in the printing rate of spheroids. Besides, given the success and efficiency of the print head in picking up and transferring a full layer of spheroids to the printing surface, we can quickly increase the number of holes in the bottom of the print head to accommodate more spheroids and thus produce larger tissues.

Compared to other methods which utilize single-spheroid printing, a process requiring a multitude of precise positioning controls and visual analysis, this layer-by-layer method allows to straightforwardly pick up and place the spheroids [22]. This not only leads to a faster printing speed but also decreases the overall cost of our device, which totaled approximately $2000, a significant reduction in price compared to the only commercially available option.

Despite the advancements in scaffold-free spheroid printing that have been shown here, there remain a few limitations in this presented design. One in particular is the inability to customize the size and shape of the tissues. Future iterations will need to have the ability to print different tissue shapes (e.g., vessel, round, etc.) to expand the capabilities to other fields of tissue engineering. Also, the current design is limited to holding only one type of spheroid for printing. Having the choice to print layers of different types of spheroids would increase the customization potential. Further design changes could allow for the placement of multiple spheroid reservoirs on the spheroid bath stage of the bioprinter.

5. Conclusions

In conclusion, we present here a custom device for printing spheroids using a novel layer-by-layer method. This bioprinter can efficiently pick up and transfer spheroids to the printing stage in a single, complete layer, decreasing the printing time of large tissue constructs. With this precise and accurate transfer of spheroid-like beads to the printing surface, custom tissues can easily be constructed on the needles in future experiments. Also, we were able to prove, using hiPSC spheroids, that we can successfully print a single layer of spheroids onto the needle array printing surface. In addition, we were able to build and manufacture this device at an affordable price, making it easier for other researchers to use the technology if brought to the market. Although there are additional design considerations that can be incorporated into future iterations, this device helps to advance and improve the growing area of scaffold-free bioprinting and could one day become a go-to method in tissue engineering.

Supplementary Materials: The following are available online at http://www.mdpi.com/2072-666X/10/9/570/s1, Video S1: Layer-by-layer bioprinting process. Video S2: Alginate bead synthesis. Video S3: Printing of alginate beads onto the needle array. Video S4: Printing of two layers of alginate beads onto the needle array. Spreadsheet S1: Major data for figures including efficiency data, colocalization data, and alginate bead diameter data. Image S1: Immunolabeling of hiPSC spheroids.

Author Contributions: Conceptualization, J.Z.; Methodology, W.L., A.M., A.M.K.-K., and J.Z.; Software, W.L.; Validation, W.L., A.M., D.P., and A.M.K.-K.; Formal Analysis, W.L.; Investigation, W.L., A.M., D.P., and A.M.K.-K.; Resources, J.Z.; Data Curation, W.L.; Writing—Original Draft Preparation, W.L.; Writing—Review & Editing, W.L., A.M., D.P., A.M.K.-K., R.K. and J.Z.; Visualization, W.L., D.P., A.M.K.-K., R.K., and J.Z.; Supervision, R.K. and J.Z.; Project Administration, J.Z.; Funding Acquisition, J.Z.

Funding: This research was funded by the National Heart, Lung, and Blood Institute (NHLBI); grant numbers 5R01HL114120, 5R01HL131017, 5U01HL134764 to J.Z. Also, additional funding was provided by the American Heart Association through the Predoctoral Fellowship; grant number: 19PRE34380484 to W.L. and the Scientist Development Grant; grant number: 17SDG33670677 to R.K. The funders had no role in study design, data collection and analysis, decision to publish, or preparation of the manuscript.

Conflicts of Interest: The authors declare no conflict of interest.

References

1. Alonzo, M.; AnilKumara, S.; Romana, B.; Tasnima, N.; Joddar, B. 3D Bioprinting of cardiac tissue and cardiac stem cell therapy. *Transl. Res.* **2019**, *211*, 64–83. [CrossRef] [PubMed]
2. Oliveira, E.P.; Malysz-Cymborskad, I.; Golubczykd, D.; Kalkowskid, L.; Kwiatkowskad, J.; Reisabc, R.L.; Oliveiraabc, J.M.; Walczak, P. Advances in bioinks and in vivo imaging of biomaterials for CNS applications. *Acta Biomater.* **2019**, *95*, 60–72. [CrossRef] [PubMed]
3. Cidonio, G.; Glinka, M.; Dawson, J.I.; Oreffo, R.O.C. The cell in the ink: Improving biofabrication by printing stem cells for skeletal regenerative medicine. *Biomaterials* **2019**, *209*, 10–24. [CrossRef]
4. Derr, K.; Zou, J.; Luo, K.; Song, M.J.; Sittampalam, G.S.; Zhou, C.; Michael, S.; Ferrer, M.; Derr, P. Fully 3D Bioprinted Skin Equivalent Constructs with Validated Morphology and Barrier Function. *Tissue Eng. Part C Methods* **2019**, *25*, 334–343. [CrossRef] [PubMed]
5. Das, S.; Kim, S.-W.; Choi, Y.-J.; Lee, S.; Lee, S.-H.; Kong, J.-S.; Park, H.-J.; Cho, H.-W.; Jang, J. Decellularized extracellular matrix bioinks and the external stimuli to enhance cardiac tissue development in vitro. *Acta Biomater.* **2019**, *95*, 188–200. [CrossRef] [PubMed]
6. Wang, X.; Ao, Q.; Tian, X.; Fan, J.; Tong, H.; Hou, W.; Bai, S. Gelatin-Based Hydrogels for Organ 3D Bioprinting. *Polymers* **2017**, *9*, 401. [CrossRef] [PubMed]
7. Kačarević, Ž.P.; Rider, P.M.; Alkildani, S.; Retnasingh, S.; Smeets, R.; Jung, O.; Ivanišević, Z.; Barbeck, M. An Introduction to 3D Bioprinting: Possibilities, Challenges and Future Aspects. *Materials* **2018**, *11*, 2199. [CrossRef]
8. Amaral, A.J.R.; Pasparakis, G. Cell membrane engineering with synthetic materials: Applications in cell spheroids, cellular glues and microtissue formation. *Acta Biomater.* **2019**, *90*, 21–36. [CrossRef]
9. Zhang, L.; Yang, G.; Johnson, B.N.; Jia, X. Three-dimensional (3D) printed scaffold and material selection for bone repair. *Acta Biomater.* **2019**, *84*, 16–33. [CrossRef]
10. Mondschein, R.J.; Kanitkar, A.; Williams, C.B.; Verbridge, S.S.; Long, T.E. Polymer structure-property requirements for stereolithographic 3D printing of soft tissue engineering scaffolds. *Biomaterials* **2017**, *140*, 170–188. [CrossRef]
11. Liu, F.; Chen, Q.; Liu, C.; Ao, Q.; Tian, X.; Fan, J.; Tong, H.; Wang, X. Natural Polymers for Organ 3D Bioprinting. *Polymers* **2018**, *10*, 1278. [CrossRef] [PubMed]

12. Rodriguez, M.J.; Dixon, T.A.; Cohen, E.; Huang, W.; Omenetto, F.G.; Kaplan, D.L. 3D freeform printing of silk fibroin. *Acta Biomater.* **2018**, *71*, 379–387. [CrossRef] [PubMed]
13. Ji, S.; Almeida, E.; Guvendiren, M. 3D bioprinting of complex channels within cell-laden hydrogels. *Acta Biomater.* **2019**. [CrossRef] [PubMed]
14. Kiyotake, E.A.; Douglas, A.W.; Thomas, E.E.; Nimmo, S.L.; Detamore, M.S. Development and quantitative characterization of the precursor rheology of hyaluronic acid hydrogels for bioprinting. *Acta Biomater.* **2019**, *95*, 176–187. [CrossRef] [PubMed]
15. Maiti, B.; Díaz Díaz, D. 3D Printed Polymeric Hydrogels for Nerve Regeneration. *Polymers* **2018**, *10*, 1041. [CrossRef] [PubMed]
16. Kim, M.-H.; Takeuchi, K.; Kino-oka, M. Role of cell-secreted extracellular matrix formation in aggregate formation and stability of human induced pluripotent stem cells in suspension culture. *J. Biosci. Bioeng.* **2019**, *127*, 372–380. [CrossRef]
17. Schell, J.Y.; Wilks, B.T.; Patel, M.; Franck, C.; Chalivendra, V.; Cao, X.; Shenoy, V.B.; Morgan, J.R. Harnessing cellular-derived forces in self-assembled microtissues to control the synthesis and alignment of ECM. *Biomaterials* **2016**, *77*, 120–129. [CrossRef] [PubMed]
18. Desroches, B.R.; Zhang, P.; Choi, B.R.; King, M.E.; Maldonado, A.E.; Li, W.; Rago, A.; Liu, G.; Nath, N.; Hartmann, K.M.; et al. Functional scaffold-free 3-D cardiac microtissues: A novel model for the investigation of heart cells. *Am. J. Physiol. Heart Circ. Physiol.* **2012**, *302*, H2031–H2042. [CrossRef]
19. Kunz-Schughart, L.A.; Schroeder, J.A.; Wondrak, M.; van Rey, F.; Lehle, K.; Hofstaedter, F.; Wheatley, D.N. Potential of fibroblasts to regulate the formation of three-dimensional vessel-like structures from endothelial cells in vitro. *Am. J. Physiol. Heart Circ. Physiol.* **2006**, *290*, C1385–C1398. [CrossRef]
20. Mironov, V.; Visconti, R.P.; Kasyanov, V.; Forgacs, G.; Drake, C.J.; Markwald, R.R. Organ printing: Tissue spheroids as building blocks. *Biomaterials* **2009**, *30*, 2164–2174. [CrossRef]
21. Norotte, C.; Marga, F.S.; Niklason, L.E.; Forgacs, G. Scaffold-free vascular tissue engineering using bioprinting. *Biomaterials* **2009**, *30*, 5910–5917. [CrossRef] [PubMed]
22. Moldovan, N.I.; Hibino, N.; Nakayama, K. Principles of the Kenzan Method for Robotic Cell Spheroid-Based Three-Dimensional Bioprinting. *Tissue Eng. Part B Rev.* **2016**, *23*, 237–244. [CrossRef] [PubMed]
23. Arai, K.; Murata, D.; Verissimo, A.R.; Mukae, Y.; Itoh, M.; Nakamura, A.; Morita, S.; Nakayama, K. Fabrication of scaffold-free tubular cardiac constructs using a Bio-3D printer. *PLoS ONE* **2018**, *13*, e0209162. [CrossRef] [PubMed]
24. Ong, C.S.; Fukunishi, T.; Zhang, H.; Huang, C.Y.; Nashed, A.; Blazeski, A.; DiSilvestre, D.; Vricella, L.; Conte, J.; Tung, L.; et al. Biomaterial-Free Three-Dimensional Bioprinting of Cardiac Tissue using Human Induced Pluripotent Stem Cell Derived Cardiomyocytes. *Sci. Rep.* **2017**, *7*, 4566. [CrossRef] [PubMed]
25. Kizawa, H.; Nagao, E.; Shimamura, M.; Zhang, G.; Torii, H. Scaffold-free 3D bio-printed human liver tissue stably maintains metabolic functions useful for drug discovery. *Biochem. Biophys. Rep.* **2017**, *10*, 186–191. [CrossRef] [PubMed]
26. Itoh, M.; Nakayama, K.; Noguchi, R.; Kamohara, K.; Furukawa, K.; Uchihashi, K.; Toda, S.; Oyama, J.; Node, K.; Morita, S. Scaffold-Free Tubular Tissues Created by a Bio-3D Printer Undergo Remodeling and Endothelialization when Implanted in Rat Aortae. *PLoS ONE* **2015**, *10*, e0136681. [CrossRef]
27. Mattapally, S.; Zhu, W.; Fast, V.G.; Gao, L.; Worley, C.; Kannappan, R.; Borovjagin, A.; Zhang, J. Spheroids of cardiomyocytes derived from human induced-pluripotent stem cells improve recovery from myocardial injury in mice. *Am. J. Physiol.* **2018**, *315*, H327–H339. [CrossRef] [PubMed]
28. Segale, L.; Giovannelli, L.; Mannina, P.; Pattarino, F. Calcium Alginate and Calcium Alginate-Chitosan Beads Containing Celecoxib Solubilized in a Self-Emulsifying Phase. *Scientifica* **2016**, *2016*, 8. [CrossRef]
29. Olmer, R.; Lange, A.; Selzer, S.; Kasper, C.; Haverich, A.; Martin, U.; Zweigerdt, R. Suspension culture of human pluripotent stem cells in controlled, stirred bioreactors. *Tissue Eng. Part C Methods* **2012**, *18*, 772–784. [CrossRef]
30. Vanderburgh, J.; Sterling, J.A.; Guelcher, S.A. 3D Printing of Tissue Engineered Constructs for In Vitro Modeling of Disease Progression and Drug Screening. *Ann. Biomed. Eng.* **2017**, *45*, 164–179. [CrossRef]

31. Langhans, S.A. Three-Dimensional in Vitro Cell Culture Models in Drug Discovery and Drug Repositioning. *Front. Pharmacol.* **2018**, *9*, 6. [CrossRef] [PubMed]
32. Borovjagin Anton, V.; Ogle Brenda, M.; Berry Joel, L.; Zhang, J. From Microscale Devices to 3D Printing. *Circ. Res.* **2017**, *120*, 150–165. [CrossRef] [PubMed]
33. Cui, H.; Nowicki, M.; Fisher, J.P.; Zhang, L.G. 3D Bioprinting for Organ Regeneration. *Adv. Healthc. Mater.* **2017**, *6*. [CrossRef] [PubMed]

Article

A Novel Biodegradable Multilayered Bioengineered Vascular Construct with a Curved Structure and Multi-Branches

Yuanyuan Liu [1], Yi Zhang [1], Weijian Jiang [2], Yan Peng [1,*], Jun Luo [1], Shaorong Xie [1], Songyi Zhong [1], Huayan Pu [1], Na Liu [1] and Tao Yue [1,*]

1 School of Mechatronic Engineering and Automation, Shanghai University, Shanghai 200444, China; yuanyuan_liu@shu.edu.cn (Y.L.); zhangyishu@shu.edu.cn (Y.Z.); luojun@shu.edu.cn (J.L.); srxie@shu.edu.cn (S.X.); zhongsongyi@shu.edu.cn (S.Z.); phygood_2001@shu.edu.cn (H.P.); liuna_sia@shu.edu.cn (N.L.)

2 School of Mechanical Engineering, Shanghai Jiao Tong University, Shanghai 200240, China; weijiandjiang@163.com

* Correspondence: pengyan@shu.edu.cn (Y.P.); tao_yue@shu.edu.cn (T.Y.); Tel.: +86-021-66136396 (Y.P.); +86-021-66136396 (T.Y.)

Received: 15 March 2019; Accepted: 21 April 2019; Published: 24 April 2019

Abstract: Constructing tissue engineered vascular grafts (TEVG) is of great significance for cardiovascular research. However, most of the fabrication techniques are unable to construct TEVG with a bifurcated and curved structure. This paper presents multilayered biodegradable TEVGs with a curved structure and multi-branches. The technique combined 3D printed molds and casting hydrogel and sacrificial material to create vessel-mimicking constructs with customizable structural parameters. Compared with other fabrication methods, the proposed technique can create more native-like 3D geometries. The diameter and wall thickness of the fabricated constructs can be independently controlled, providing a feasible approach for TEVG construction. Enzymatically-crosslinked gelatin was used as the material of the constructs. The mechanical properties and thermostability of the constructs were evaluated. Fluid-structure interaction simulations were conducted to examine the displacement of the construct's wall when blood flows through it. Human umbilical vein endothelial cells (HUVECs) were seeded on the inner channel of the constructs and cultured for 72 h. The cell morphology was assessed. The results showed that the proposed technique had good application potentials, and will hopefully provide a novel technological approach for constructing integrated vasculature for tissue engineering.

Keywords: bioengineered vascular constructs; curved structure; multi-branches; enzymatically-crosslinked; tissue engineering

---

1. Introduction

In the field of tissue engineering, one methodology that is often used is to integrate signal factors, cells and biomaterial constructs to replace or improve functional tissues [1,2]. In this common practice, biomaterial constructs act as the basic structure to support and guide cell growth under the influence of some biologically active molecules [3]. An ideal biomimetic structure must have properties (e.g., biocompatibility, elasticity, extensibility) similar to those of the extracellular matrix (ECM) of the native tissue [4]. The selection of the structure and the materials of the biomimetic construct is crucial for effectively mimicking the ECM of human body.

Coronary bifurcation is vulnerable to atherosclerosis as a result of the distinct local blood flow patterns [5] caused by the bifurcated structure of the coronary artery [6]. The blood flow through the artery exerts shear stress onto this particular area, which makes this area susceptible to lesions [7], and which subsequently leads to the coronary artery bifurcation lesion [8], one common lesion of the coronary artery. The treatment of coronary artery bifurcation lesions is a major challenge for interventional cardiology [9]. There exist several interventional techniques for the treatment of this disease, such as the kissing balloon dilatation, crush technique, amongst others [10]. However, the operation process of these techniques remains complex, and the complication rates are usually high. The results of the interventional techniques are usually not satisfactory [11]. For example, the bifurcation area is often subjected to occlusion after operation, and the restenosis rate is very high. Instead of employing traditional interventional techniques, the development of a substitution of the coronary arteries for transplantation is a new method that holds great promise [12].

As far as the treatment of vascular diseases such as the abovementioned coronary bifurcation lesions is concerned, autografts are an ideal candidate in transplantation operations [13]. They possess a benign biocompatibility to the host patients, and the rejection rates are usually low. However, the number of autografts is often rare because of the lack of suitable donors and harvest sites [14]. Synthetic grafts composed of polytetrafluoroethylene and Dacron perform well in replacing large-diameter blood vessels (inner diameter > 6 mm), but they are not suitable for small-diameter conditions [15]. In the channel of small-diameter vessels, the blood flow rate is low and the shear stress is high, which often causes thrombosis and neointimal hyperplasia. In contrast, scaffolds made of natural materials such as gelatin and chitosan often perform better [16].

Some of the recent tissue engineered vascular graft (TEVG) fabricating methods tend to mimic the native blood vessels, which feature a multilayered structure, multi-branches and a spatial configuration [17–19]. Like the vessels in vivo, the multilayered structure is helpful for aligning different types of cells, such as endothelial cells, fibroblasts, and smooth muscle cells, in different layers and improving the interaction between them [20], which is a tremendous progress in better mimicking the structure of a native blood vessel. Additionally, in human tissue, the number of vessel branches increases while their inner diameters decrease [21]. This morphology could ensure the mass transportation within the whole vascular network. Therefore, when designing the relatively large-diameter portion of a vessel network, the branching factor must be taken into consideration. Finally, for the purpose of realizing an even distribution of the vessel network in a three-dimensional space, the configuration of the large-diameter channel should be designed as a three-dimensional structure rather than a planar one. However, due to the limitations of the fabrication techniques and materials, it still currently remains a challenge to produce bioengineered vascular constructs that meet all of the abovementioned characteristics.

Among materials commonly used for biological tissue fabrication, gelatin is a natural polymer similar to collagen; it possesses benign biocompatibility and biodegradability [22,23]. Due to these properties, gelatin could be an ideal candidate for cell delivery and could serve as the substitute for ECM. Some recent studies have used microbial transglutaminase (mTG) to crosslink gelatin to increase the biocompatibility and mechanical strength of engineered constructs [24–26]. Enzymatically crosslinked gelatin gels were proven to be biocompatible with human cells. For this reason, a gelatin/mTG biopolymer was usually chosen as the ECM-like material for fabricating biomimetic constructs. Besides, Pluronic F127, which is composed of polyoxyethylene (PEO) and polyoxypropylene (PPO), undergoes thermally reversible gelation above a lower critical solution temperature (LCST) [27,28]. Therefore, it is suitable as the fugitive ink when constructing a hollow channel.

In the present study, a novel additive and subtractive hybrid manufacturing technique combining 3D printed molds, casting hydrogel and sacrificial material was proposed. Compared to current fabrication techniques, the proposed method fabricated multilayered TEVG with a curved structure and multi-branches for the first time, to our best knowledge. The material of the construct was enzymatically crosslinked gelatin. The vascular geometry design was based on the model of the left coronary artery (LCA), hence the fabricated vessel-like construct possessed an interconnected channel whose diameter varied in the range of 1.8 to 3 mm. The proposed technique is a promising low-cost approach for tissue engineering.

2. Materials and Methods

2.1. Materials

Gelatin (Type A, 300 bloom from porcine skin) and Pluronic F127 was purchased from Sigma Co. Ltd. (St. Louis, MO, USA). Microbial transglutaminase (mTG) was purchased from Ajinomoto Inc. (Kanagawa, Japan); its activity was approximately 100 U/g.

Gelatin was dissolved in deionized water and was continuously stirred in a 60 °C water bath for 30 min. When the temperature of the gelatin solution dropped to 37 °C, mTG was added and thoroughly blended with the gelatin solution. The gelatin/mTG solution contains 14% (wt) gelatin and 1.4% (wt) mTG. The fugitive ink was composed of 40% (wt) Pluronic F127 in deionized water. The ink was homogenized using a mixer until the powder was fully dissolved, and then centrifuged to remove any air bubbles. The fugitive ink was subsequently loaded in a syringe and stored at 4 °C.

2.2. Geometry of the Tissue Engineered Vascular Grafts (TEVG)

The left coronary artery (LCA) consists of the left main artery (LM) that bifurcates into the left anterior descending (LAD) and the left circumflex (LCX) [29]. The constructed geometry model is presented in Figure 1. The wall thicknesses of the middle layer and the outermost layer are set to be 0.5 mm and 0.5 mm, respectively. The innermost layer was formed by seeding human umbilical vein endothelial cells (HUVECs) on the inner wall of the channel. The dimensions of the proposed TEVG geometry are based on the clinically accurate dimensions of the LCA. The dimensions of the inner channel of the TEVG are presented in Table 1 [30]. The tapering effects were considered when constructing the LAD and LCX. The angle of 60° was used between the two branches [6]. The curvature was set to be 60 mm.

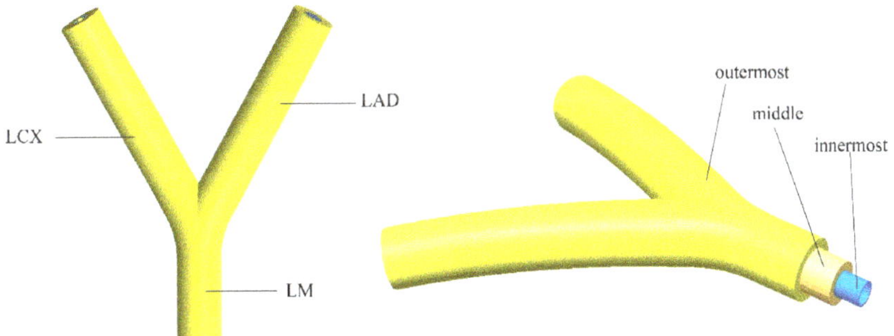

Figure 1. The geometry model of the proposed tissue engineered vascular grafts (TEVGs) based on left coronary artery. LCX–left circumflex; LAD–left anterior descending; LM–left main artery.

Table 1. Dimensions of the inner channel of the tissue engineered vascular grafts (TEVGs).

Vessel	Diameter (mm)		Length (mm)	Curvature Radius (mm)
	Inlet	Outlet		
Left main artery (LM)	1.3	1.3	6	60
Left circumflex (LCX)	1.3	1.1	15	60
Left anterior descending (LAD)	1.3	1.1	15	60

2.3. Fluid-Structure Interaction Simulation

The strength of a fabricated TEVG under native blood pressure is crucial for its application in the clinic. To investigate the property of the wall compliance of the constructed TEVG under the influence of the blood flow, a fluid-structure interaction (FSI) simulation was performed via COMSOL Multiphysics. The geometrical model is presented in Figure 2a along with the mesh model used in the simulation (Figure 2b). The engineered vascular construct is embedded in biological tissue, specifically the cardiac muscle. The fluid domain was also constructed. The flowing blood applies pressure to the internal surfaces, thereby deforming the vascular construct and the cardiac muscle.

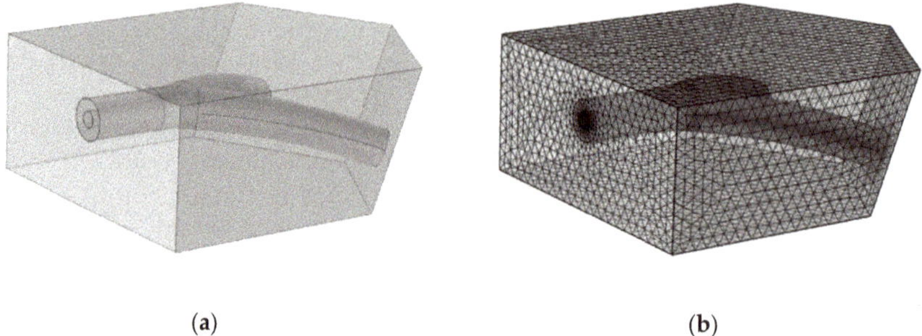

(**a**) (**b**)

Figure 2. (**a**) The geometry of the bioengineered vascular construct embedded in the cardiac muscle used for the simulations, and (**b**) the mesh model created for the simulation.

The simulation consists of two distinct but coupled studies: first, a fluid-dynamics study of the blood; second, a mechanical study of the deformation of the bioengineered vascular construct and cardiac muscle. To correctly estimate the deformation response of the TEVG, the mechanical study must consider the cardiac muscle because it exerts a stiffness that resists the vascular construct deformation due to the applied pressure.

Blood was modeled as a Newtonian fluid with the use of the incompressible Navier-Stokes equations; as for LAC dimensions, the shear rate is well over the limit where blood exhibits shear-thinning behavior for the cardiac cycle. A laminar flow model was chosen as the calculated Reynolds number was below 100. The density and viscosity of the blood were set to be 1050 kg/m^3 and 4.0×10^{-3} Pa·s [31]. No slip boundary condition was imposed for the inner walls of the construct. An incompressible Neo-Hookean solid model was chosen for the TEVG and cardiac muscle so as to better predict its nonlinear stress-strain behavior. The Poisson's ratio (υ) of the TEVG and cardiac muscle was set to be 0.45 [32,33]. As for the Lamé's coefficients of the neo-Hookean hyperelastic materials, the Lamé's second coefficients (μ) of the TEVG and muscle were set to be 6.20×10^6 N/m^2 and 7.20×10^6 N/m^2, respectively. Therefore, the Lamé's first coefficients (λ) could be calculated by the following equations [34]:

$$\lambda = K - \frac{2}{3}G \tag{1}$$

$$\upsilon = \frac{3K - 2G}{6K + 2G} \tag{2}$$

$$\mu = G \tag{3}$$

where K and G represent the bulk modulus and shear modulus of the hyperelastic materials, respectively.

During a heart beating cycle, the pressure varies between a minimal and a maximal value [35]. Therefore, for the time-dependent analysis, a simplified waveform of the pressure for the inlet boundary condition of the fluid domain was used and is shown in Figure 3. The maximal pressure value occurs at the time t = 0.5 s. The pressure between 0 and 1 s makes the pressure vary between the minimal and maximal value during a heart beating cycle. The pressure variation during a cardiac cycle is within the range of the normotensive pulse pressure.

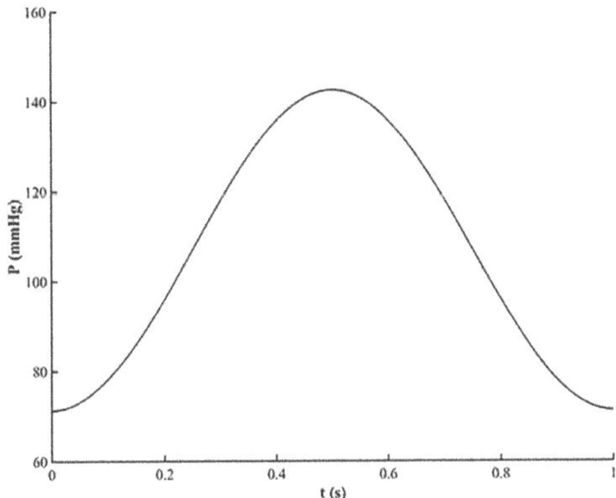

Figure 3. The time variation of the pressure used as the inlet boundary condition for the TEVG.

Additionally, the compliance of the vascular construct can be calculated as follows:

$$\%\text{compliance} = \frac{(R_{p_2} - R_{p_1})/R_{p_1}}{(p_2 - p_1)} \times 10^4 \quad (4)$$

where

p_1 is the lower pressure value (mmHg);
p_2 is the higher pressure value (mmHg);
R_{p_1} is the internal radius at the lower pressure value (mm);
R_{p_2} is the internal radius at the higher pressure value (mm).

2.4. Fabrication Method

2.4.1. Fabrication of Mold System

The methods employed for the fabrication of the outermost layer and middle layer of the TEVG were realized via a mold system. The mold system consists of five separate molds. They were made of photosensitive resin (Clear Resin, Formlabs, Somerville, MA, USA) and printed directly by a 3D printer (Form 2, Formlabs) from stereolithography (STL) files (SolidWorks 2018, Dassault Systèmes, Vélizy-Villacoublay, France). The printing parameters were set following the general configurations in the program. Briefly, the layer thickness was set to be 0.025 mm and the resin temperature was 35 °C. In the settings of the supporting constructions, the density of the supports, the contact point size and the thickness of the basement were set to be 1.00, 0.7 mm and 2 mm respectively. The profiles of the five molds are shown in Figure 4. The cross section of each mold's track is semicircular. The center trajectories of each mold's track are identical. The curvature radiuses of the fitting surfaces of the five molds are also the same (60 mm), which corresponds to the curved structure of the bioengineered TEVG. The diameter of the groove/convex on each mold's curved surface is determined by the dimensions of the designed TEVG.

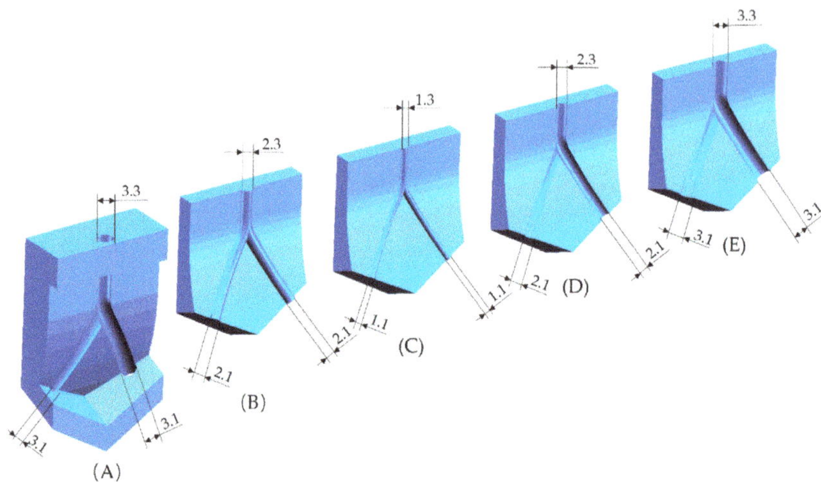

Figure 4. Mold system: (**A**) mold 1; (**B**) mold 2; (**C**) mold 3; (**D**) mold 4; and (**E**) mold 5.

2.4.2. Fabrication of Multilayered Bifurcated TEVG with Curved Structure

A schematic representation of the step-by-step process for the fabrication of multilayered bifurcated TEVGs with a curved structure is presented in Figure 5a. At first, mold 1 and mold 2 were fitted together, with their center trajectories coincident, and three plastic tubes were inserted into the flank hole of mold 1 as the mold side gate to compose the feed system. Next, a freshly prepared gelatin/mTG solution was housed in a syringe and pipetted into the channel formed by mold 1 and mold 2 (Figure 5b). The solution delivery rate was maintained at 6 mL/min by a micro-pump attached to the syringe. After cooling at room temperature for 15 min and then 30 min at 10 °C to induce the gelation of the gelatin injected in the molds, mold 2 was gently removed, leaving the hydrogel structure in the channel of mold 1, which corresponds to half of the outermost layer of the TEVG. In the same manner, mold 3 was fitted together with mold 1, and the gelatin/mTG solution was pipetted into the channel from the gate, gelatinating to form half of the middle layer of the structure. After the gelation of the gelatin, we removed mold 3. Then, the fugitive ink Pluronic F127 was loaded in a syringe and printed into the groove of the hydrogel on mold 1. The printing process was executed by a custom-built three-axle linkage platform (Figure 5c). The diameter of the printed F127 ink could be regulated by adjusting the moving speed of the platform and the delivery rate of the micro-pump online. After the printing process, the rest was done in the same manner as mentioned above. Briefly, mold 4 and mold 5 were subsequently fitted together with mold 1, and the gelatin/mTG solution was pipetted into the inner channel to shape the left half of the middle layer and outermost layer of the TEVG respectively. After being crosslinked at room temperature for 15 min and then at 10 °C for 30 min, the upper mold was gently removed; then, we carefully removed mold 1, and a multilayered vessel-like structure was achieved. We placed the construct at 4 °C for 20 min, and during this process, fugitive ink Pluronic F127 liquefied and flowed away, forming the inner channel of the TEVG. The obtained construct was incubated at 37 °C for 5 h to get fully crosslinked. Then, the construct was immersed in 65 °C distilled water for 5 h to heat-inactivate the residual enzyme.

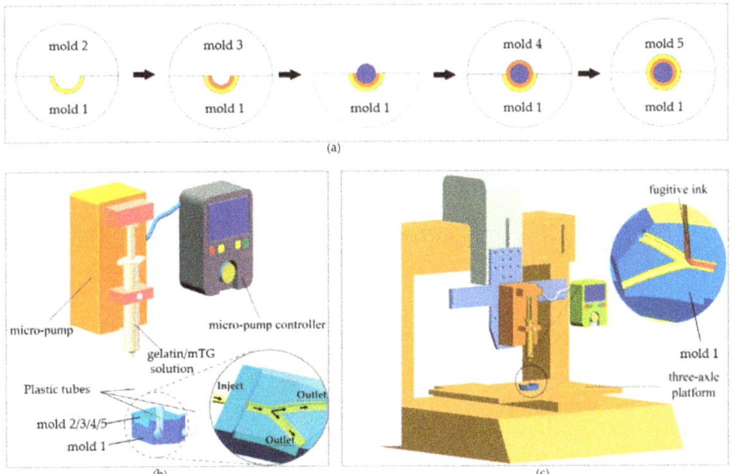

Figure 5. Fabrication process of the TEVG: (**a**) procedure of the step-by-step fabrication process. The molds 2–5 were gathered after mold 1 to compose the injection mold systems to form the different layers of the TEVG; (**b**) the gelatin/mTG perfusion system; (**c**) the fugitive ink printing system.

2.5. In Vitro Cytocompatibility of the TEVG

Human umbilical cord derived endothelial cells (HUVECs) were trypsinized, centrifuged and suspended in a culture medium at a cell density of 1.0×10^6 cells/mL. The cells in the third passage after defrosting were used for the experiments. The TEVG was soaked in 75% ethanol for 1 h and then washed 3 times with phosphate buffer saline (PBS) to remove the ethanol. After that, the sample was exposed to ultraviolet light for 30 min. The cell suspension was injected into the channel. After 4 h of cell attachment, the cellular construct was placed in Dulbecco's modified Eagle medium containing high glucose and sodium pyruvate (DMEM) (HyClone, GE), supplemented with 10% fetal bovine serum (FBS) (HyClone, GE), and cultured in a humidified incubator at 37 °C. It was statically cultured with medium that was changed each day. At 4 h and 72 h after culturing, the TEVG was washed in sterile PBS (HyClone, GE) three times and stained using Live-Dye^{TM} ("live"; 1 µL/mL; BioVision) and propidium iodide ("dead"; 1 µL/mL; BioVision) for 30 min. After that, the cellular morphology and fluorescent images were observed with an inversed fluorescent microscope (Eclipse Ti-U, Nikon Instruments Inc., Tokyo, Japan).

2.6. Uniaxial Compressive Testing

The uniaxial compressive testing was operated on a material testing machine (Z2.5, ZWICK, Ulm, Germany). To ensure stability during the test, the diameter of the cross-section was scaled up by three times its original diameter. The uniaxial compressive tests were performed at the rate of 1 mm/min at an ambient temperature of 23 °C and 37 °C respectively. The environmental humidity was about 65%. To investigate the effect of the addition of mTG on the mechanical properties, samples with and without crosslinking were fabricated to carry out the tests.

2.7. Thermostability of the TEVG

To examine the thermostability of the bioengineered TEVG, samples with and without mTG were both submerged in PBS (pH 7.4) and stored in the incubator at 37 °C for different time periods (2, 4, 6, 8 and 10 days). After each time period, the samples were washed and subsequently dried in a vacuum oven at room temperature for 6 h. Their weights were recorded during this period.

3. Results

3.1. Fluid-Structure Interaction Simulation

The mechanical properties of biomimetic constructs under the effect of actual human body conditions are critical in terms of the clinical applications and potential failure. In the human body, blood vessels usually undergo very large blood pressure, especially in the case of a large-diameter vessel. Biomaterials such as gelatin usually undergo very large strains under loading, and the stress-strain relationship is generally nonlinear. The behavior of the TEVG is influenced not only by its geometry and material properties, but also by the fluid passing through it. Numerical simulations were carried out to estimate the displacement of the wall of the TEVG under the influence of the blood flow.

The inclusion of surrounding cardiac tissue is also essential for the FSI simulation process, because during the expansion process of the TEVG's wall the cardiac muscle would exert a tethering effect on the wall and thus restrict the radial motion. Hence, the integrated simulation model is more desirable than the one consisting solely of the TEVG, thus providing more realistic boundary conditions for the simulation.

The total displacement of the TEVG and cardiac muscle at the peak pressure (t = 0.5 s) is shown in Figure 6a,b, respectively. The maximal displacement of 1.18 μm occurs at the bifurcated location. The displacement in the bifurcated area is large since this is the location where the blood flow impinges on the wall of the construct, leading to a local high-pressure region. No rupture was observed from the results, implying that the risk of failure is low under coronary blood flow conditions. The compliance of the vascular construct is calculated to be 0.26%, which is within the normal range of a human blood vessel [36].

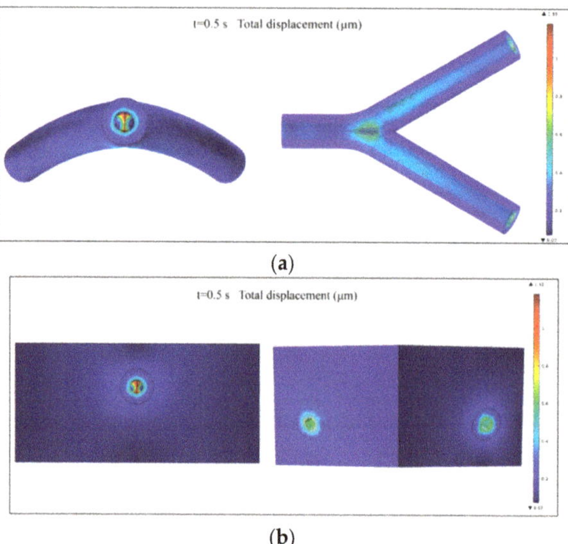

(b)

Figure 6. (**a**) The total displacement of the TEVG at the peak load (t = 0.5 s); and (**b**) the total displacement of the whole construct at the peak load (t = 0.5 s).

Additionally, for the time-dependent analysis, the total displacements of the TEVG in a complete heart beating cycle (0–1 s) are shown in Figure 7a–c. It could be observed that the peak displacement of the structure is still mainly located in the bifurcated region. The simulation results can be used to examine the risk of failure for the fabricated TEVG when it is integrated in other three-dimensional tissue constructs.

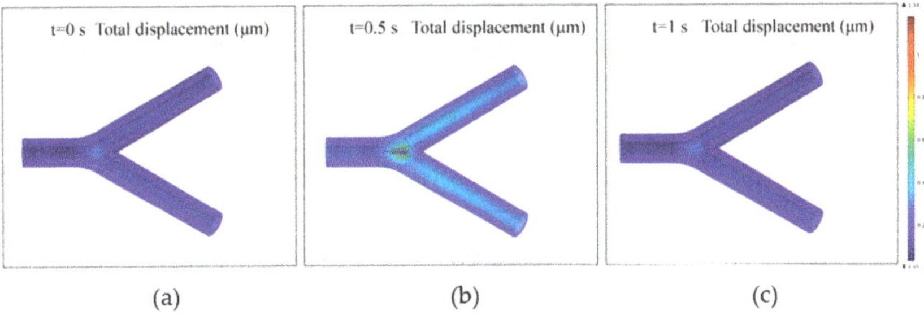

Figure 7. The time variation of the total displacement of the TEVG at: (**a**) t = 0 s; (**b**) t = 0.5 s; and (**c**) t = 1 s.

3.2. The Morphology of the Proposed TEVG

In this study, a novel method of fabricating biodegradable multilayered TEVGs was proposed. The fabricated enzymatically-crosslinked gelatin vascular construct is shown in Figure 8. The top view (Figure 8a) of the construct confirmed its bifurcated structure in the three-dimensional space. To reveal the layered structure of the construct, blue and red acrylic paints were added in the hydrogel of each layer when fabricating the construct. It can be observed that the TEVG possesses a distinct multilayered structure (Figure 8b, blue: outermost layer of the structure; and red: middle layer of the structure). The inner channel of the construct became visible after being immersed in water (Figure 8c), which demonstrated its connectivity. The section view of the TEVG revealed that the molding generated a real circular shape after sacrificing the F-127 (Figure 8d) and form a bifurcated channel ready for liquid perfusion (Video S1). The innermost layer partially deviated from the center, which was supposed to be affected by the die clearance. Compared with previously seen TEVGs, the proposed TEVGs possess a branched structure while keeping the multilayered feature. This is of particular importance because over-simplified in vitro TEVG models have a limited significance for simulating the actual in vivo environment.

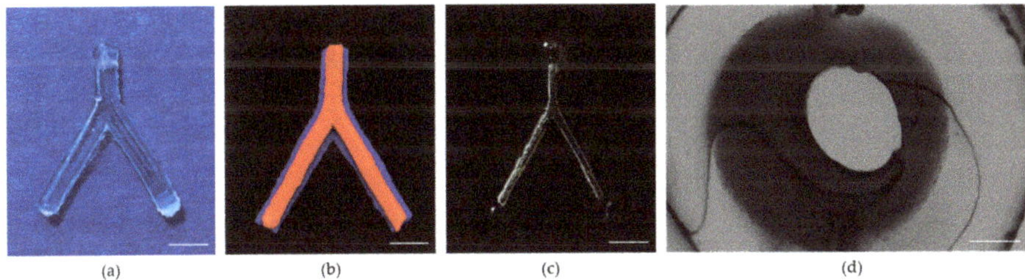

Figure 8. Morphology of the TEVG: (**a**) top view; (**b**) multilayered structure of the TEVG (blue: outermost layer of the structure; and red: middle layer of the structure); (**c**) inner channel of the construct after being submerged in water. (Scale bars: 5 mm); and (**d**) section view of the TEVG (Scale bar: 500 μm).

3.3. The Results after Cells Seeding

One of the most critical properties of biomimetic structures are their in vitro cytocompatibility, and the creation of a continuous monolayer of ECs in the inner channel is necessary for the proper function of TEVGs [37]. To investigate the in vitro cytocompatibility of the fabricated TEVGs, HUVECs were seeded and statically cultured in the inner channel. The microscopic morphology image (Figure 9a,c) and the fluorescent image (Figure 9b,d) of endothelial cells after 4 h of attachment were shown. It could be observed that the HUVECs adhere well on the lumina. However, the cell distribution at this stage is quite nonuniform, and this might be caused by the uneven distribution of the cell in the suspension at the initial stage. Additionally, because the inner diameter of the channel is quite large, the adhesion of the cells is difficult. Many cells entering from the inlet didn't have the chance to adhere and were flushed away from the outlet. After 72 h of culturing, as could be observed from Figure 9e–h, the amount of HUVECs had significantly increased, and the cells in the inner channel were found to have well spread on the wall and to have taken a normal cellular phenotype. In spite of the nonuniformity of the distribution in the attachment phase, the cells still covered the surface of the channel and formed an endothelialized monolayer. The results indicated that the crosslinked TEVG had a good in vitro cytocompatibility and was suitable for cell attachment and for proliferation.

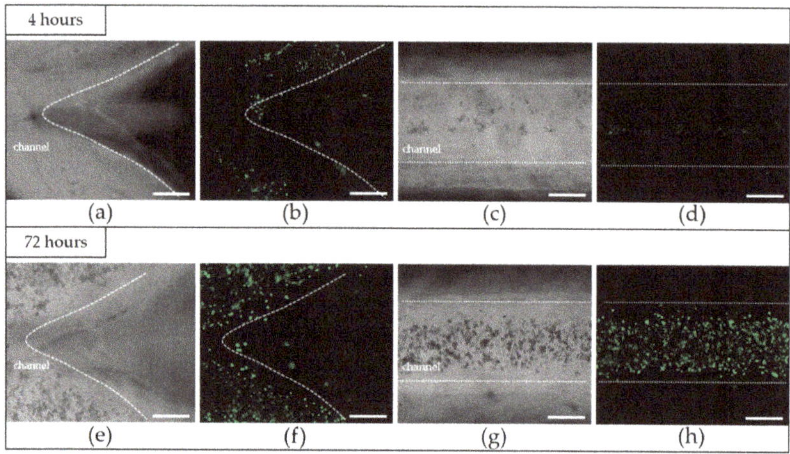

Figure 9. Microscopic images and fluorescent images of cells; the dotted lines were used to divide the inside and outside of the channel: (**a**,**b**) after 4 h of attachment at the bifurcate area of the TEVG, (**c**,**d**) after 4 h of attachment at the linear channel of the TEVG, (**e**,**f**) after 72 h of culturing at the bifurcate area of the TEVG, (**g**,**h**) after 72 h of culturing at the linear channel of the TEVG. (Scale bars: 500 μm).

3.4. Mechanical Properties of the Proposed TEVG

The mechanical properties of tissue-engineered biological constructs determine their application fields. The TEVGs fabricated in this study possess a high aspect ratio, which determines their susceptibility to bucking when suffering from a compressive load. Therefore, the uniaxial compressive mechanical properties of the fabricated TEVG under room temperature and physiological conditions were investigated. The mechanical properties of gelatin samples with and without crosslinking were evaluated. Figure 10a shows the compressive stress-strain curves for up to a 60% strain. The compressive modulus of the four groups of samples are shown in Figure 10b. It can be observed that the addition of mTG (1.4%, wt) increased the compressive modulus of the constructs by more than three times, from 1.5 MPa to 5 MPa at room temperature. Under 37 °C, the mechanical properties of the gelatin/mTG TEVG is almost the same as that under 23 °C, while the pure gelatin one completely solved. The main reason for this mechanical enhancement lies in the fact that the composition of gelatin includes glutamine and lysine. The presence of these amino acids means there are more crosslinking sites in gelatin, thus promoting the chance of creating stiffer gels when provided with sufficient amounts of mTG under proper crosslinking conditions. Compared with the TEVG composed of GelMA hydrogels, the maximum compressive modulus of the gelatin/mTG based TEVG is enhanced more than ten-fold [38].

The mechanical properties of the enzymatically-crosslinked TEVG are still relatively low compared to those of native blood vessels. However, as was envisioned, the real value of the fabricated vessel-mimic structure lies in its capacity to carry various types of vascular cells and to eventually develop into function blood vessels. Hence, the capability of maintaining its original morphology when integrated with other engineered scaffolds is sufficient for the proper function of the fabricated TEVG.

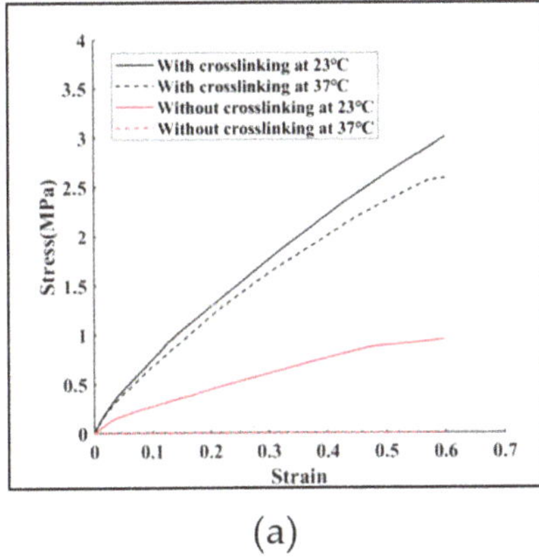

(a)

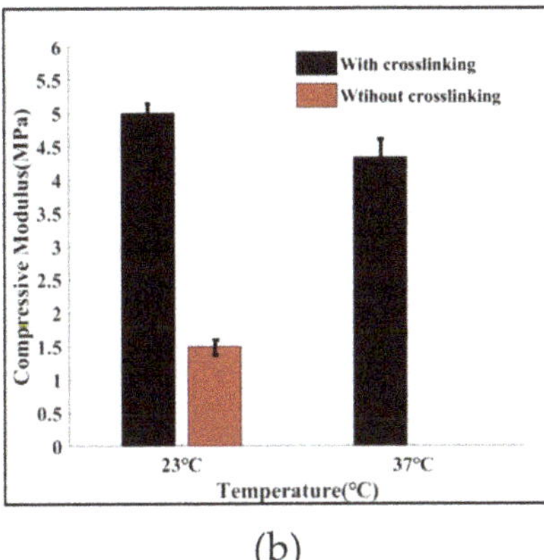

(b)

Figure 10. Compressive testing. (**a**) stress-strain curves of the samples with or without crosslinking at 23 °C and 37 °C respectively; and (**b**) compressive modulus of the samples with or without crosslinking at 23 °C and 37 °C respectively.

3.5. Thermostability

One possible application of the fabricated TEVG is its potential of being integrated into other three-dimensional scaffolds. As the carrier of different types of vascular cells, the designed TEVG has the potential of developing into a functional multilayered blood vessel. One crucial premise is the ability of maintaining its design morphology under the temperature of the human body. Figure 11 shows the in vitro degradation process of the gelatin samples with and without crosslinking. It could be observed that the enzymatically-crosslinked TEVGs degrade much slower than those without crosslinking, with a degradation rate of 16.54%. In contrast, the samples composed purely of gelatin completely dissolved within two days. One possible explanation for the enhancement is that the addition of mTG increased the intermolecular association through the formation of covalent bonds in the gel matrix, thus dramatically slowing down the degradation speed of the gelatin construction. The results indicate that the gelatin/mTG biopolymer possesses a distinct advantage over the pure gelatin polymer in terms of thermostability, which proved the feasibility of using the crosslinked construct as the carrier of cells to form a functional vessel.

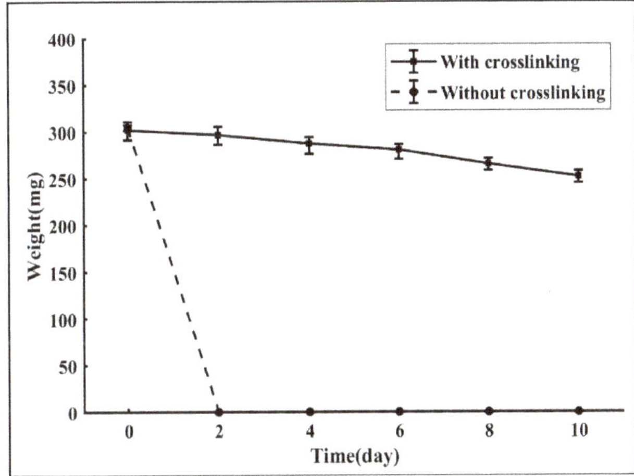

Figure 11. In vitro degradation of the TEVG with and without crosslinking.

4. Discussion

The biological compositions of each layer of the native vessels are different. Likewise, the wall thicknesses and mechanical properties of different layers also vary. For instance, the wall thickness of the vessels near the heart are relatively larger. Consequently, when fabricating TEVGs, the ability to independently control each layer's composition and thickness is essential in order to achieve their biological and mechanical performances such as the biocompatibility and the compressive strength.

Today, extrusion-based printing [39], inkjet printing [40], stereolithography [41], two-photon polymerization [42] and laser-assisted printing [43] have been used wisely in vasculature engineering. While the above methods successfully shape tissue-engineered vessels having a multilayered structure or achieve vascularization in bulk constructions, there remains the challenge of fabricating curved bifurcated vascular constructions with a multilayered wall at the macro-scale in a controllable way. In the current study, a novel method of fabricating multilayered biodegradable TEVGs with a curved structure and multi-branches was proposed. The mold system was designed by computer aided design (CAD) software and fabricated by a 3D printer. Taking advantage of the merits of 3D printing technology, we can create three-dimensional structures that are closer to the native blood vessel's morphologies. The proposed method also allows the variation of the composition and dimension of each layer of the constructs. By varying the design parameters of each layer, for example, or by possibly incorporating various bioactive substances in each layer, the thickness and biocompatibility of each layer can be independently controlled. Furthermore, the fabricated tissue-engineered TEVGs possessed a bifurcated structure, which was determined by the CAD models of the mold system. Thus, it is also possible to fabricate vascular constructs with more branches by simply modifying the CAD models. In addition, the three-dimensional structure of the constructs could also be easily modified by varying the design parameters of the mold system. Because of this design flexibility, 3D printed mold systems will give researchers the ability to quickly and inexpensively design TEVGs with the shape and complexity they desire.

In recent years, several methodologies have been introduced to create channels in bulk hydrogels. Bertassoni et al. embedded agarose stripes in in the hydrogel construction to form the inner networks [44]. Negrini et al. sacrificed alginate templates through a chelating agent to obtain a porous gelatin hydrogel [45]. Li et al. added a pre-printed bifurcated polyvinyl alcohol (PVA) structure to the hydrogel to realize vascularization [46]. Compared with the fugitive materials listed above, the adoption of the Pluronic F127 fugitive ink as the sacrificial material is an easy way of forming the inner channel, as the Pluronic F127 is a thermal sensitive material. Furthermore, by taking advantage of the strong shear thinning response of F127 at room temperature, F127 can be printed into very complex channel networks and maintains its shape fidelity without harmful chemical cross-linking, heat processing or additional templates preparing to fabricate inner networks. For TEVGs below the LCST, the ink liquefies and flows readily, while at the same time the other hydrogel materials used are stiff and solid-like, which makes it easily removed by temperature variation without being obstructed by the closed structure and without exerting extra mechanical force that may impair the printed structure. Taking advantage of this complimentary behavior, the printed ink could be removed without influencing the form of other materials. Our work successfully solved the problem of creating multiple branches for a multi-layered TEVG, which means that the fabricated TEVGs possess more features similar to the native blood vessel. This is a big step towards creating TEVGs with the same level of complexity, which has a tremendous significance for in vitro cardiovascular research and could hopefully reduce the demand for animal experiments.

HUVECs were found to adhere well to the luminal surface of the sample. The formed continuous monolayer of HUVECs was evident after 72 h, which demonstrated the good biocompatibility of the enzymatically-crosslinked TEVGs and their potential for acting as a substitute for ECM.

From the acquired results, one can figure out that the proposed triple-layered TEVG could easily be imbedded into other porous tissue-engineered scaffolds, thus forming a multi-scale vasculature within a whole three-dimensional structure; the scale of the whole vasculature could range from micrometers to millimeters. The existence of this multi-scale vasculature will greatly facilitate the mass transport within the scaffold. Furthermore, when combined with techniques such as cell encapsulation, the fabricated construct would have the potential of developing into a real functional blood vessel. Specifically, if the cell-encapsulating hydrogels were used for the fabrication of the TEVG, different types of vascular cells would further develop into different layers of the blood vessel with the degradation of the hydrogel materials.

As for the scale of the proposed construct, whether the smallest inner diameter could be achieved by this technique depends to a certain degree on the resolution of the 3D printing technology. Theoretically speaking, by promoting the accuracy of the 3D printer, or by adopting techniques with a higher precision such as micro-fluids technology, the scale range of the inner diameter of the TEVG could be further expanded, thus widening the application field of this novel process. Hence, this fabrication technique holds great potential for impacting a wide range of fields.

Finally, another promising advantage of this novel technique lies in its capacity to design patient specific vessel models because the mold can be constructed by a 3D printer which could be driven directly by the computed tomography (CT) scan data of patients' original vessels, which means that personalized TEVGs that are more suited to the actual human body conditions can be constructed. Constructing the mold system from the clinical data of native vessels is an interesting point for future exploration, and this personalized diagnostic approach has tremendous potential in clinical diagnoses and treatments.

5. Conclusions

In this study, a novel approach for fabricating a multilayered biodegradable bioengineered vascular construct with a cured structure and multi-branches was presented. The method used in this study combined 3D printed molds, casting hydrogel and sacrificial material, providing an easy technique to construct TEVGs whose morphology and structure are close to those of the original blood vessels. With the inherent advantages of 3D printing, it is envisioned that the proposed technique will play a significant role in the field of tissue engineering and in assisting studies on cardiovascular diseases.

Supplementary Materials: The following are available online at http://www.mdpi.com/2072-666X/10/4/275/s1, Video S1: Liquid perfusion inside the channels.

Author Contributions: Investigation, Y.L., Y.Z., W.J. and S.Z.; methodology, Y.L., Y.Z., H.P. and Y.P.; writing—original draft preparation, Z.Y., W.J.; writing—review and editing, Y.L., Y.P., J.L. and S.X.; validation, N.L., T.Y.

Funding: This research was funded by National Natural Science Foundation of China (grant number 51475281, 51575333) and China National Funds for Distinguished Young Scientists (grant number 61625304).

Conflicts of Interest: The authors declare no conflict of interest.

References

1. Jiang, L.Y.; Li, Y.; Xiong, C.D.; Su, S.P.; Ding, H.J. Preparation and Properties of Bamboo Fiber/Nano-hydroxyapatite/Poly(lactic-co-glycolic) Composite Scaffold for Bone Tissue Engineering. *ACS Appl. Mater. Interfaces* **2017**, *9*, 4890–4897. [CrossRef] [PubMed]
2. Levorson, E.J.; Kasper, F.K.; Mikos, A.G. 5.501 - Scaffolds: Flow Perfusion Bioreactor Design. *Compre. Biomater.* **2011**, *5*, 1–11.
3. Rajzer, I.; Menaszek, E.; Castano, O. Electrospun polymer scaffolds modified with drugs for tissue engineering. *Mater. Sci. Eng. C* **2017**, *77*, 493–499. [CrossRef] [PubMed]
4. Stocco, T.D.; Rodrigues, B.V.M.; Marciano, F.R.; Lobo, A.O. Design of a novel electrospinning setup for the fabrication of biomimetic scaffolds for meniscus tissue engineering applications. *Mater. Lett.* **2017**, *196*, 221–224. [CrossRef]
5. Zuin, M.; Rigatelli, G. Treatment of de Novo Coronary Artery Bifurcation Lesions with Drug Coated Balloons: A Reappraisal According to the Available Scientific Data. *Cardiovasc. Revasc. Med.* **2018**, *19*, 57–64. [CrossRef] [PubMed]

6. Givehchi, S.; Safari, M.J.; Tan, S.K.; Shah, M.N.B.M.; Sani, F.B.M.; Azman, R.R.; Sun, Z.; Yeong, C.H.; Ng, K.H.; Wong, J.H.D. Measurement of coronary bifurcation angle with coronary CT angiography: A phantom study. *Phys. Med.* **2018**, *45*, 198–204. [CrossRef] [PubMed]
7. Redfors, B.R.; Généreux, P. Percutaneous Coronary Intervention for Bifurcation Lesions. *Interv. Cardiol. Clin.* **2016**, *5*, 153–175. [CrossRef] [PubMed]
8. Doutel, E.; Carneiro, J.; Campos, J.B.L.M.; Miranda, J.M. Experimental and numerical methodology to analyze flows in a coronary bifurcation. *Eur. J. Mech. B Fluids* **2018**, *67*, 341–356. [CrossRef]
9. Leesar, M.A.; Hakeem, A.; Azarnoush, K.; Thuesen, L. Coronary bifurcation lesions: Present status and future perspectives. *Int. J. Cardiol.* **2015**, *187*, 48–57. [CrossRef] [PubMed]
10. Vinayakumar, D.; Uppalakal, B.; Goyal, K.K.; Nair, A. Balloon embedded stenting: A novel technique for percutaneous coronary intervention of bifurcation lesions, experience in Indian population. *Indian Heart J.* **2018**, *70*, 278–281. [CrossRef] [PubMed]
11. He, X.; Gao, B.; Liu, Y.; Li, Z.; Zeng, H. Side-Branch Technique for Difficult Guidewire Placement in Coronary Bifurcation Lesion. *Cardiovasc. Revasc. Med.* **2016**, *17*, 59–62. [CrossRef]
12. Kawamoto, H.; Ruparelia, N.; Tanaka, A.; Chieffo, A.; Latib, A.; Colombo, A. Bioresorbable Scaffolds for the Management of Coronary Bifurcation Lesions. *JACC Cardiovasc. Int.* **2016**, *9*, 989–1000. [CrossRef] [PubMed]
13. Cahueque, M.; Macias, D.; Moreno, G. Reconstruction with non-vascularized fibular autograft after resection of clavicular benign tumor. *J. Orthop.* **2015**, *12*, S255–S259. [CrossRef] [PubMed]
14. Matsumura, G.; Hibino, N.; Ikada, Y.; Kurosawa, H.; Biomaterials, T. Successful application of tissue engineered vascular autografts: Clinical experience. *Biomaterials* **2003**, *24*, 2303–2308. [CrossRef]
15. Agaimy, A.; Ben-Izhak, O.; Lorey, T.; Scharpf, M.; Rubin, B.P. Angiosarcoma arising in association with vascular Dacron grafts and orthopedic joint prostheses: Clinicopathologic, immunohistochemical, and molecular study. *Ann. Diagn. Pathol.* **2016**, *21*, 21–28. [CrossRef] [PubMed]
16. Choi, Y.R.; Kim, E.H.; Lim, S.; Choi, Y.S. Efficient preparation of a permanent chitosan/gelatin hydrogel using an acid-tolerant tyrosinase. *Biochem. Eng. J.* **2018**, *129*, 50–56. [CrossRef]
17. Song, W.; Lu, Y.C.; Frankel, A.S.; An, D.; Schwartz, R.E.; Ma, M. Engraftment of human induced pluripotent stem cell-derived hepatocytes in immunocompetent mice via 3D co-aggregation and encapsulation. *Sci. Rep.* **2015**, *5*, 16884. [CrossRef] [PubMed]
18. Hikosaka, M.; Murata, A.; Yoshino, M.; Hayashi, S.-I. Correlation between cell aggregation and antibody production in IgE-producing plasma cells. *Biochem. Biophys. Rep.* **2017**, *10*, 224–231. [CrossRef] [PubMed]
19. Hasan, A.; Paul, A.; Memic, A.; Khademhosseini, A. A multilayered microfluidic blood vessel-like structure. *Biomed. Microdev.* **2015**, *17*. [CrossRef] [PubMed]
20. Wolf, F.; Vogt, F.; Schmitz-Rode, T.; Jockenhoevel, S.; Mela, P. Bioengineered Vascular Constructs as Living Models for In Vitro Cardiovascular Research. *Drug Discovery Today* **2016**, *21*, 1446–1455. [CrossRef] [PubMed]
21. Hasan, A.; Paul, A.; Vrana, N.E.; Zhao, X.; Memic, A.; Hwang, Y.S.; Dokmeci, M.R.; Khademhosseini, A. Microfluidic techniques for development of 3D vascularized tissue. *Biomaterials* **2014**, *35*, 7308–7325. [CrossRef] [PubMed]
22. Alizadeh, M.; Abbasi, F.; Khoshfetrat, A.B.; Ghaleh, H. Microstructure and characteristic properties of gelatin/chitosan scaffold prepared by a combined freeze-drying/leaching method. *Mater. Sci. Eng. C* **2013**, *33*, 3958–3967. [CrossRef] [PubMed]
23. Wu, X.; Liu, Y.; Li, X.; Wen, P.; Zhang, Y.; Long, Y.; Wang, X.; Guo, Y.; Xing, F.; Gao, J. Preparation of aligned porous gelatin scaffolds by unidirectional freeze-drying method. *Acta Biomater.* **2010**, *6*, 1167–1177. [CrossRef]
24. Paguirigan, A.L.; Beebe, D.J. Protocol for the fabrication of enzymatically crosslinked gelatin microchannels for microfluidic cell culture. *Nat. Protoc.* **2007**, *2*, 1782–1788. [CrossRef] [PubMed]
25. He, J.; Chen, R.; Lu, Y.; Zhan, L.; Liu, Y.; Li, D.; Jin, Z. Fabrication of circular microfluidic network in enzymatically-crosslinked gelatin hydrogel. *Mater. Sci. Eng. C* **2016**, *59*, 53–60. [CrossRef] [PubMed]
26. Liu, F.; Majeed, H.; Antoniou, J.; Li, Y.; Ma, Y.; Yokoyama, W.; Ma, J.G.; Zhong, F. Tailoring physical properties of transglutaminase-modified gelatin films by varying drying temperature. *Food Hydrocoll.* **2016**, *58*, 20–28. [CrossRef]
27. Yap, L.S.; Yang, M.-C. Evaluation of Hydrogel Composing of Pluronic F127 and Carboxymethyl Hexanoyl Chitosan as Injectable Scaffold for Tissue Engineering Applications. *Colloids Surf. B* **2016**, *146*, 204–211. [CrossRef] [PubMed]
28. Kolesky, D.B.; Truby, R.L.; Gladman, A.S.; Busbee, T.A.; Homan, K.A.; Lewis, J.A. 3D Bioprinting of Vascularized, Heterogeneous Cell-Laden Tissue Constructs. *Adv. Mater.* **2014**, *26*, 3124–3130. [CrossRef]

29. Mizuguchi, Y.; Hashimoto, S.; Shibutani, H.; Yamada, T.; Taniguchi, N.; Nakajima, S.; Hata, T.; Takahashi, A. Successful Treatment of a Nonagenarian Patient with Acute Coronary Syndrome Complicated with Chronic Total Occlusion of the Left Main Coronary Artery. *Cardiovasc. Revasc. Med.* **2016**, *18*, 276–280. [CrossRef]
30. Dodge, J.T., Jr.; Brown, B.G.; Bolson, E.L.; Dodge, H.T. Lumen diameter of normal human coronary arteries. Influence of age, sex, anatomic variation, and left ventricular hypertrophy or dilation. *Circulation* **1992**, *86*, 232–246. [CrossRef]
31. Torii, R.; Wood, N.B.; Hadjiloizou, N.; Dowsey, A.W.; Wright, A.R.; Hughes, A.D.; Davies, J.; Francis, D.P.; Mayet, J.; Yang, G.-Z.; et al. Fluid–structure interaction analysis of a patient-specific right coronary artery with physiological velocity and pressure waveforms. *Int. J. Numer. Methods Biomed. Eng.* **2009**, *25*, 565–580. [CrossRef]
32. Yousefi-Banaem, H.; Kermani, S.; Asiaei, S.; Sanei, H. Prediction of myocardial infarction by assessing regional cardiac wall in CMR images through active mesh modeling. *Comput. Biol. Med.* **2017**, *80*, 56–64. [CrossRef] [PubMed]
33. Weiss, A.V.; Fischer, T.; Iturri, J.; Benitez, R.; Toca-Herrera, J.L.; Schneider, M. Mechanical properties of gelatin nanoparticles in dependency of crosslinking time and storage. *Colloids Surf. B* **2019**, *175*, 713–720. [CrossRef] [PubMed]
34. Jadamba, B.; Khan, A.A.; Raciti, F. On the inverse problem of identifying Lamé coefficients in linear elasticity. *Comput. Math. Appl.* **2008**, *56*, 431–443. [CrossRef]
35. Katritsis, D.G.; Theodorakakos, A.; Pantos, I.; Andriotis, A.; Efstathopoulos, E.P.; Siontis, G.; Karcanias, N.; Redwood, S.; Gavaises, M. Vortex formation and recirculation zones in left anterior descending artery stenoses: Computational fluid dynamics analysis. *Phys. Med. Biol.* **2010**, *55*, 1395–1411. [CrossRef] [PubMed]
36. Sarkar, S.; Salacinski, H.J.; Hamilton, G.; Seifalian, A.M. The mechanical properties of infrainguinal vascular bypass grafts: Their role in influencing patency. *Eur. J. Vasc. Endovasc.* **2006**, *31*, 627–636. [CrossRef]
37. Zhang, C.; Hu, K.; Liu, X.; Reynolds, M.A.; Bao, C.; Wang, P.; Zhao, L.; Xu, H.H.K. Novel hiPSC-based tri-culture for pre-vascularization of calcium phosphate scaffold to enhance bone and vessel formation. *Mater. Sci. Eng. C* **2017**, *79*, 296–304. [CrossRef]
38. Nichol, J.W.; Koshy, S.T.; Bae, H.; Hwang, C.M.; Yamanlar, S.; Khademhosseini, A. Cell-laden microengineered gelatin methacrylate hydrogels. *Biomaterials* **2010**, *31*, 5536–5544. [CrossRef]
39. Gao, Q.; He, Y.; Fu, J.Z.; Liu, A.; Ma, L. Coaxial nozzle-assisted 3D bioprinting with built-in microchannels for nutrients delivery. *Biomaterials* **2015**, *61*, 203–215. [CrossRef] [PubMed]
40. Christensen, K.; Xu, C.X.; Chai, W.X.; Zhang, Z.Y.; Fu, J.Z.; Huang, Y. Freeform Inkjet Printing of Cellular Structures with Bifurcations. *Biotechnol. Bioeng.* **2015**, *112*, 1047–1055. [CrossRef]
41. Huber, B.; Engelhardt, S.; Meyer, W.; Kruger, H.; Wenz, A.; Schonhaar, V.; Tovar, G.E.M.; Kluger, P.J.; Borchers, K. Blood-Vessel Mimicking Structures by Stereolithographic Fabrication of Small Porous Tubes Using Cytocompatible Polyacrylate Elastomers, Biofunctionalization and Endothelialization. *J. Funct. Biomater.* **2016**, *7*, 11. [CrossRef]
42. Meyer, W.; Engelhardt, S.; Novosel, E.; Elling, B.; Wegener, M.; Kruger, H. Soft Polymers for Building up Small and Smallest Blood Supplying Systems by Stereolithography. *J. Funct. Biomater.* **2012**, *3*, 257–268. [CrossRef]
43. Guillotin, B.; Souquet, A.; Catros, S.; Duocastella, M.; Pippenger, B.; Bellance, S.; Bareille, R.; Remy, M.; Bordenave, L.; Amedee, J.; et al. Laser assisted bioprinting of engineered tissue with high cell density and microscale organization. *Biomaterials* **2010**, *31*, 7250–7256. [CrossRef]
44. Bertassoni, L.E.; Cecconi, M.; Manoharan, V.; Nikkhah, M.; Hjortnaes, J.; Cristino, A.L.; Barabaschi, G.; Demarchi, D.; Dokmeci, M.R.; Yang, Y.Z.; et al. Hydrogel bioprinted microchannel networks for vascularization of tissue engineering constructs. *Lab Chip* **2014**, *14*, 2202–2211. [CrossRef] [PubMed]
45. Negrini, N.C.; Bonnetier, M.; Giatsidis, G.; Orgill, D.P.; Fare, S.; Marelli, B. Tissue-mimicking gelatin scaffolds by alginate sacrificial templates for adipose tissue engineering. *Acta Biomater.* **2019**, *87*, 61–75. [CrossRef]
46. Li, S.; Liu, Y.Y.; Liu, L.J.; Hu, Q.X. A Versatile Method for Fabricating Tissue Engineering Scaffolds with a Three Dimensional Channel for Prevasculature Networks. *ACS Appl. Mater. Interfaces* **2016**, *8*, 25096–25103. [CrossRef]

 *micromachines*

Review

Advanced Polymers for Three-Dimensional (3D) Organ Bioprinting

Xiaohong Wang [1,2]

1 Center of 3D Printing & Organ Manufacturing, School of Fundamental Sciences, China Medical University (CMU), No. 77 Puhe Road, Shenyang North New Area, Shenyang 110122, China; wangxiaohong709@163.com or wangxiaohong@tsinghua.edu.cn; Tel./Fax: +86-24-31900983

2 Center of Organ Manufacturing, Department of Mechanical Engineering, Tsinghua University, Beijing 100084, China

Received: 25 October 2019; Accepted: 19 November 2019; Published: 25 November 2019

Abstract: Three-dimensional (3D) organ bioprinting is an attractive scientific area with huge commercial profit, which could solve all the serious bottleneck problems for allograft transplantation, high-throughput drug screening, and pathological analysis. Integrating multiple heterogeneous adult cell types and/or stem cells along with other biomaterials (e.g., polymers, bioactive agents, or biomolecules) to make 3D constructs functional is one of the core issues for 3D bioprinting of bioartificial organs. Both natural and synthetic polymers play essential and ubiquitous roles for hierarchical vascular and neural network formation in 3D printed constructs based on their specific physical, chemical, biological, and physiological properties. In this article, several advanced polymers with excellent biocompatibility, biodegradability, 3D printability, and structural stability are reviewed. The challenges and perspectives of polymers for rapid manufacturing of complex organs, such as the liver, heart, kidney, lung, breast, and brain, are outlined.

Keywords: three-dimensional (3D) printing; organ manufacturing; biomaterials; polymers; stem cells

---

1. Introduction

In nature, all the complicated phenomena of life, including human organs, are the result of biochemical and biophysical changes of molecules (or materials at a molecular level). For example, small organic molecules or compounds combine to form larger polymers (macromolecules or biomacromolecules). Macromolecules arrange in specific ways to form cells with organelles inside the cell membrane (Figure 1). While homogeneous cells organize into tissues, heterogeneous cells aggregate into organs with particular physiological functions. It has taken, in some cases, thousands of years to evolve from tiny organic molecules to microcells, mesotissues, and macro-organs [1–3].

At present, organ failure is the main cause of mortality all over the world. Despite the rapid development in pharmacological, interventional, and surgical therapies during the last several decades, the only cure for organ failure is allograft organ transplantation, which is seriously limited by issues, such as donor organ shortage, life-long immune rejection, and ethical conflict [4–6].

Three-dimensional (3D) organ bioprinting is the utilization of 3D printing technologies to assemble multiple cell types or stem cells/growth factors along with other biomaterials in a layer-by-layer fashion to produce bioartificial organs that maximally imitate their natural counterparts [7–9]. Traditionally, 3D printing is named rapid prototyping (RP), solid freeform fabrication (SFF), or additive manufacturing (AM) based on the dispersion–accumulation (i.e., discrete–accumulation) principle of computer-aided manufacturing (CAM) techniques. Before 3D printing, an object can be divided into numerous two-dimensional (2D) layers with a defined thickness. These 2D layers can be piled up by selectively adding the desired materials in a highly reproductive layer-by-layer manner under the instruction of computer-aided design (CAD) models [10–13]. Patient-specific organ image data, such as computerized tomography (CT) and magnetic resonance imaging (MRI), can be easily transferred into CAD models for customized organ manufacturing with predefined geometrical shapes, biomaterial constituents, and physiological functions [14–17].

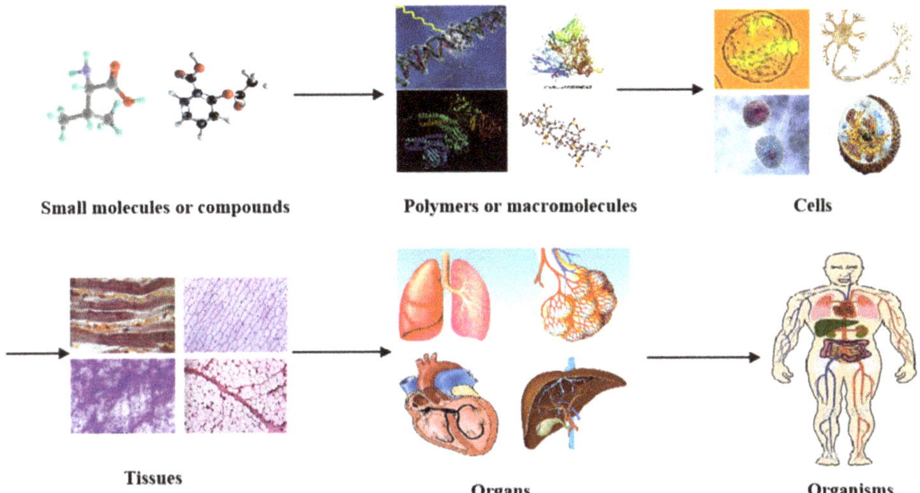

Figure 1. Different levels of materials (or molecules) existing in the human body, from small organic molecules or compounds to larger polymers (i.e., macromolecules or biomacromolecules), cells, tissues, organs, and systems in organisms.

The 3D organ bioprinting procedure involves changes to the properties of a series of materials at molecular, cell, tissue, and organ levels. It is an emerging new interdisciplinary field that needs cooperation of many fields of science and technology, such as biomaterials, biology, physics, chemistry, computers, mechanics, bioinformatics, and medicine (Figure 2).

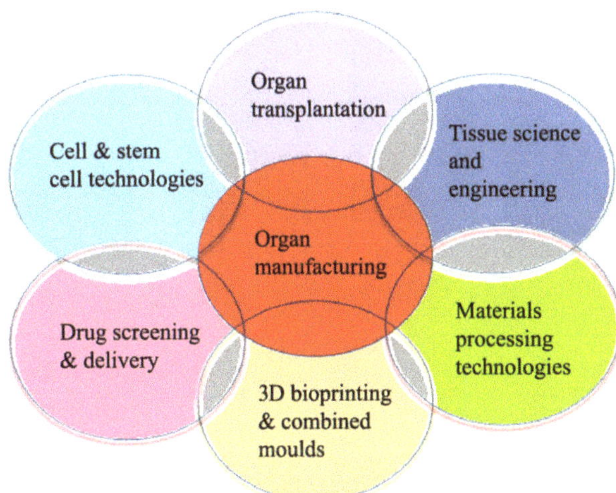

Figure 2. Relationships of organ manufacturing, including three-dimensional (3D) bioprinting, with other pertinent fields of science and technology.

During the last 16 years, a large variety of 3D bioprinting technologies have been exploited, which has led to the emergence of fully automatic manufacturing of bioartificial organs for wide biomedical applications, such as high-throughput drug screening, controlled cell transplantation, customized organ repair/regeneration/replacement/restoration, pathological mechanism analysis, metabolism model establishment, and living tissue/organ cryopreservation [10–20]. Based on the working principles, these technologies can be classified into four major groups: inkjet-based, extrusion-based, laser-based, and their combinations (Figure 3). Each of the former three groups has advantages and disadvantages in bioartificial organ manufacturing.

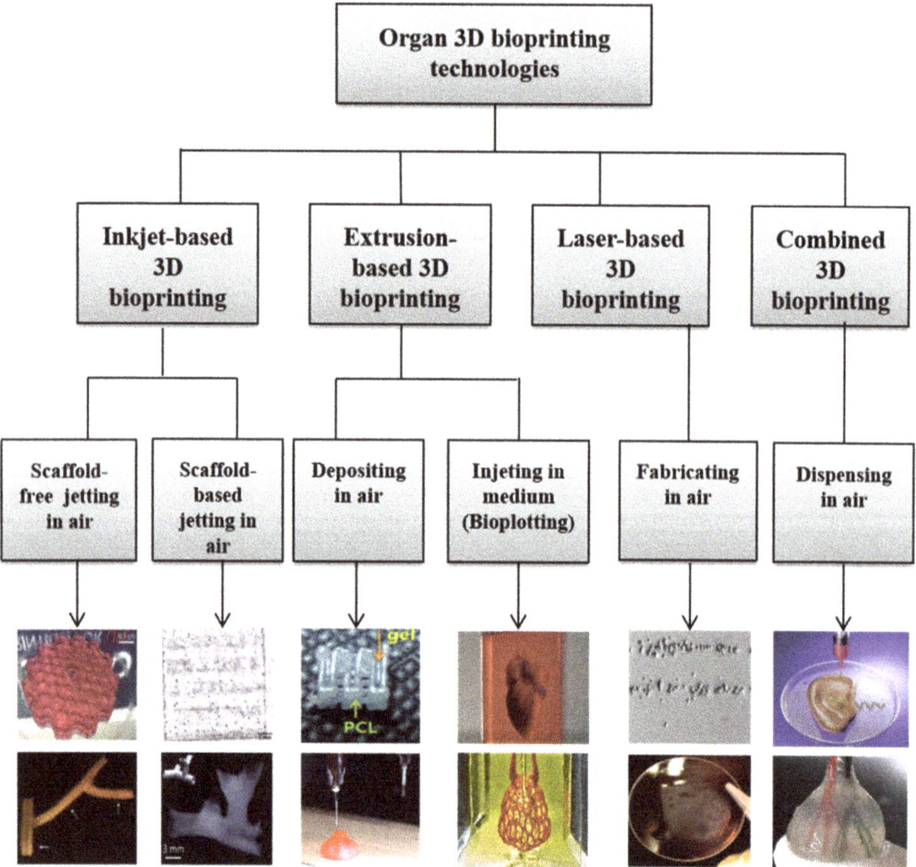

Figure 3. Graphical description of 3D bioprinting types [10–20]. Images reproduced with permission from [10–20].

Several series of unique automatic and semiautomatic bioartificial organ manufacturing technologies have been created in my own group with the proper integration of modern high technologies, including computer, biology (e.g., cells and stem cells), biomaterials (e.g., polymers), chemistry, mechanics, and medicine. With these unique high technologies, we have solved all the bottleneck problems that have perplexed tissue engineers and other scientists for more than 6–7 decades, such as large-scale tissue/organ manufacturing [21–24], hierarchical vascular/nerve network construction with a fully endothelialized inner surface and antisuture/antistress capabilities [25–29], step-by-step adipose-derived stem cell (ASC) differentiation in a 3D construct [30–32], long-term preservation of bioartificial tissues/organs [33–35], in vitro metabolism model establishment [36,37], high-throughput drug screening [38–40], in vivo biocompatibilities of implanted biomaterials [41–43]. Several polymers have played essential and ubiquitous roles for bioartificial organ manufacturing with the incorporation of multiple cell types, stem cells/growth factors, and hierarchical vascular and neural networks with antisuture and antistress functions.

2. Role of Polymers in 3D Organ Bioprinting

3D organ bioprinting is not a simple and easy engineering approach. Like building a nuclear plant, it requires intermingling of intricate architectural design, appropriate biomaterial selection, special building process, multicellular incorporation, supportive structure utilization, controllable stem cell induction, simultaneous or sequential tissue formation and maturation, multiple tissue coordination, and the means of coordinating these procedures to form large-scale living organs [25–33].

Polymers are large molecules made up of many small and identical repeating units bonded by covalent bonds [44]. The smallest repeating unit in a polymer is called a monomer, while the number of repeat units in a polymer chain is termed the degree of polymerization (DP) or chain length. Polymers can be divided into natural and synthetic groups according to their origin. Most natural polymers are water-soluble and endowed with some common biological and physiological properties, such as being pliable as soft tissues and organs, friendly for cell encapsulation and transplantation, and easy to handle and reshape. The properties of synthetic polymers depend on many factors, such as processing conditions, molecular weights, monomer distributions, chain structures (e.g., size, geometry, inter/intrabonding, and branching), and the presence of additives. In particular, linear polymers have long chains (or backbones) that contain small chemical groups on the repeating units.

Hydrogels are 3D hydrophilic networks of polymers that can absorb and retain large amounts of water and gel under certain physical (e.g., thermosensitive), chemical (e.g., covalent bonding), or biochemical (e.g., enzymatic) conditions. The physical and chemical properties of hydrogels can be designed for specific biomedical applications by selecting proper polymer components, inorganic solvents, and gelation protocols [45–47]. Compared with other states of polymers, hydrogels can provide a benign and stable environment for living cells to grow, migrate, aggregate, proliferate, and differentiate inside. The integration of hydrogels with 3D bioprinting technologies has offered numerous attractive features for complex organ manufacturing [48–50].

During the 3D organ bioprinting process, different polymers have different roles and functions. Most natural polymers, which are employed as the main components of bioinks, have the following roles: (1) provide cells and bioactive agents with support such as accommodation; (2) build vascular, neural, and lymphatic networks as semipermeatable substrates for nutrient, gas (e.g., oxygen), metabolite, and biosignal exchange; (3) guide homogeneous and heterogeneous histogenesis and organogenesis in a predefined way; and (4) promote tissue and organ maturation under specific biochemical and biophysical conditions. Meanwhile, most synthetic polymers are applied for the following functions: (1) improve multicellular handling or allotting in space to mimic their natural counterparts; (2) enhance the mechanical properties of vascular and neural networks with antisuture and antistress capabilities; (3) commit (or complete) extra functions, such as sacrificing supports and protecting covers.

Globally, the literature has reported on a number of polymers for use in 3D bioprinting. These are summarized in Figure 4 [51–74]. These polymers need to meet several basic requirements for 3D printing of bioartificial organs and subsequently clinical applications (Figure 5): (1) biocompatible (i.e., nontoxic or no obvious toxicity, no or low immunological reaction); (2) bioprintable using 3D bioprinters; (3) biostable or crosslinkable with strong enough mechanical properties; (4) biodegradable (in particular, the biodegradation rate should match the new tissue/organ generation speed); (5) suturable with host vascular and nerve networks; (6) permeable for nutrients and gases; (7) biostorable before being printed; and (8) sterilizable.

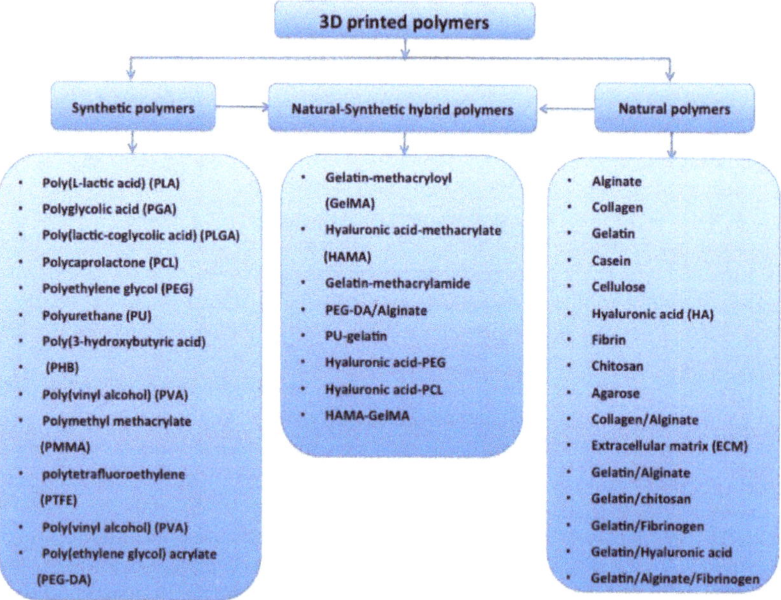

Figure 4. Polymers that have been used for tissue and 3D organ printing.

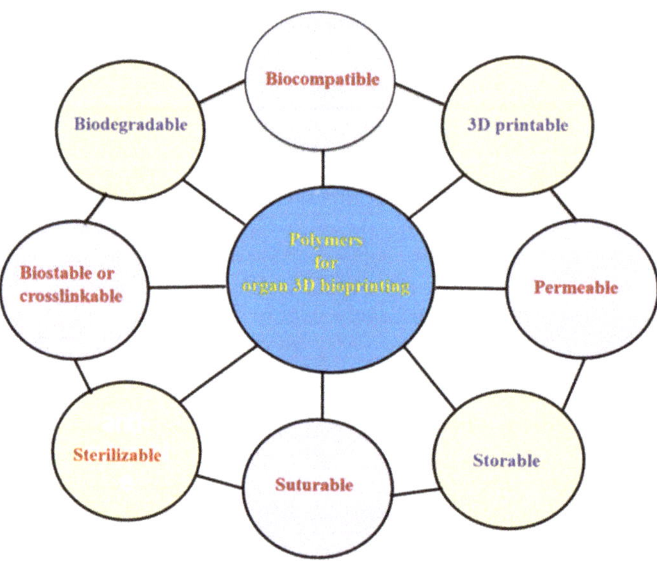

Figure 5. Basic requirements for selecting a polymer for 3D bioprinting of bioartificial organs.

In the following sections, seven normal polymers that have been frequently employed in 3D organ bioprinting with excellent biocompatibility, biodegradability, biostability, and bioprintability are individually analyzed according to their natural or synthetic properties.

3. Natural Polymers for 3D Organ Bioprinting

Natural polymers are widely existent in animal, plant, and microbe tissues as the main components of extracellular matrices (ECMs) or decellularized extracellular matrices (dECMs). These polymers include (1) proteins, such as collagen I–V, elastin, keratin, actin, tubulin, and myosin; (2) polysaccharides, such as chitin, alginate, and starch; (3) glycoproteins, such as mucin, lectin, miraculin, transferrin, and nectin; (4) proteoglycans, such as decorin, syndecan, versican, betaglycan, lumican, and fibromodulin. Compared with synthetic polymers, most natural polymers dissolve in inorganic solvents, such as water, phosphate buffer saline (PBS), and Dulbecco's modified Eagle medium (DMEM), which are cell-friendly. Few of the natural polymer solutions or hydrogels can be used directly as cell-loading matrices for 3D organ bioprinting.

Due to the special physical, chemical, and biological properties, most natural polymer solutions or hydrogels cannot be printed alone with a sol–gel transformation taking place during the 3D printing processes [25–33]. These polymers are often used as additives for several theromsensitive or chemical crosslinkable polymer (e.g., gelatin, agar/agarose, and alginate) solutions or hydrogels for 3D bioprinting. The 3D bioprintability of natural polymers are mainly determined by the molecular weight, viscosity, hydrophilicity, and crosslinkability of the polymer solutions or hydrogels. The 3D bioprinting accuracy of the cell-laden natural polymer solutions or hydrogels depends largely on the polymer concentration, viscoelasticity, gelation speed, and shear thinning behavior.

The main advantages of natural polymers for 3D organ bioprinting is that they can entrap viable cells and bioactive agents before printing, protect cells and bioactive agents during 3D printing, and form semipermeable substrates after 3D printing. Before 3D printing, cells and bioactive agents are normally embedded in natural polymer solutions or hydrogels. During 3D printing, the natural polymer chains can safeguard cells from printing stress and provide cells with predesigned 3D milieus similar to those in a native organ. After 3D printing, the polymer chains can be physically/chemically/enzymatically crosslinked to form semipermeable substrates. These semipermeable substrates are permeable to nutrients, gases (e.g., oxygen), and metabolites of cells.

At present, the most frequently used natural polymers for 3D organ bioprinting are collagen, gelatin, alginate, fibrinogen, starch, hyaluronan, chitosan, silk, dextran, agar (or agarose), and matrigel (or dECM). Among these polymers, alginate, gelatin, fibrinogen, and dECM are the most promising candidates for 3D organ bioprinting.

3.1. Alginate

Alginate, the salt of alginic acid, is an anionic polysaccharide derived from brown seaweed algae. It consists of β-D-mannuronic acid (M block) and α-L-guluronic acid (G block) monomers in its molecule. Alginate itself can dissolve in water and be crosslinked by divalent cations, such as calcium (Ca^{2+}), barium (Ba^{2+}), and strontium (Sr^{2+}) ions, due to ion exchange reactions. This characteristic is particularly attractive in many biomedical fields, such as nanoparticle preparation, drug delivery, wound healing, tissue engineering, and regenerative medicine [75–78]. An obvious characteristic of alginate solution is that its physical sol–gel transition point is below 0 °C. Under ambient temperature, it is hard for pure alginate solution to be printed in layers without chemical crosslinking. The in vivo biocompatibility of alginate is not as good as those of animal- or human-derived natural polymers, such as gelatin and fibrinogen [41].

Generally, the physiochemical properties of alginate hydrogels depend on the ratio of M/G blocks. The higher the M/G ratio, the higher the activity of the polymers. The first pertinent alginate 3D bioprinting technology was reported in 2005, in which alginate was used as an additive in gelatin-based cell-laden bioinks (Figure 6) [25–33]. The blending of alginate with gelatin molecules can improve the printing resolution and increase the shape fidelity. Only a certain range of the alginate/gelatin concentrations can be printed in layers. Optimal concentration of alginate in gelatin-based bioinks varies from 0.5% (w/v) to 3% (w/v) depending on the polymer resources. After 3D printing, calcium ion crosslinks (i.e., ion bonds) can significantly improve the structural stability to a certain degree. The 3D printed constructs can be crosslinked through various liquid exposures, such as spraying, soaking, and filtering of calcium solutions [79–81]. This exposure leads to the exchange of sodium ions in the alginate molecules with Ca^{2+} ions whereby the divalent Ca^{2+} ions form chemical crosslinks or chelates between two carboxyl groups in the same or different polymer chains.

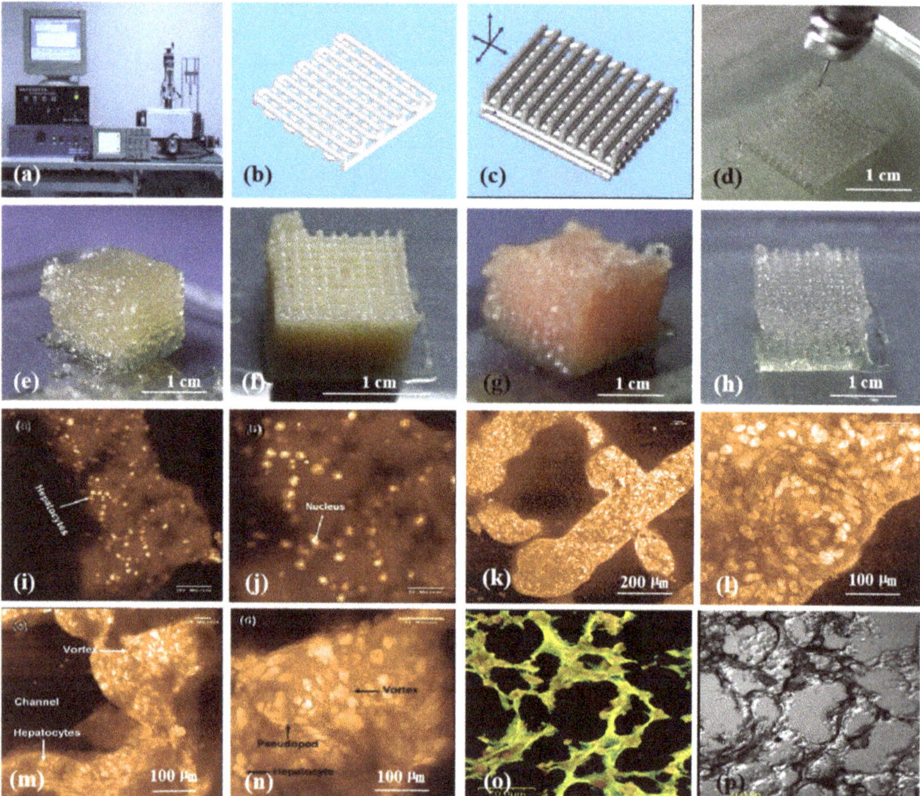

Figure 6. 3D bioprinting of chondrocytes, cardiomyocytes, hepatocytes, and adipose-derived stem cells (ASCs) into living tissues/organs using a pioneering 3D bioprinter made in Prof. Wang's laboratory at Tsinghua University: (**a**) the pioneering 3D bioprinter; (**b**) schematic description of a cell-laden gelatin-based hydrogel being printed into an one-layer grid lattice using the 3D bioprinter; (**c**) schematic description of the cell-laden gelatin-based hydrogel being printed into large-scale 3D construct using the 3D bioprinter; (**d**) 3D printing process of a chondrocyte-laden gelatin-based construct; (**e**) a grid 3D construct made from a cardiomyocyte-laden gelatin-based hydrogel; (**f**) hepatocytes encapsulated in a gelatin-based hydrogel after 3D printing; (**g**) hepatocytes in a gelatin-based hydrogel after 3D printing; (**h**) a gelatin-based hydrogel after 3D printing; (**i**) hepatocytes in a 3D printed construct after a certain period of in vitro culture; (**j**) a magnified photo of (i); (**k**) the shape of the hepatocyte-laden grid construct maintained well after certain period of in vitro culture; (**l**) a magnified photo of (k), showing hepatocytes formed vortex in the hydrogel; (**m**) hepatocytes in a 3D printed construct after a longer period of in vitro culture; (**n**) a magnified photo of (m), showing the vortex structure was still there; (**o**) immunostaining of the hepatocyte-laden 3D construct after certain period of in vitro culture, showing new hepatic tissue formed in the gelatin-based hydrogel with close cell-cell connection or tight junction; (**p**) a dark-field microscopy of (o). Images reproduced with permission from [7,22].

It is very interesting that the chemical crosslinking of alginate molecules using calcium ions is reversible. When the 3D printed constructs are placed in a liquid containing no or less Ca^{2+}, the crosslinked Ca^{2+} dissolves gradually within about one week. Further reinforcement is necessary when long-term in vitro culture is required. This means that the 3D constructs need to be further stabilized on and off using calcium ions during long-term in vitro cultures.

Alginate-based 3D bioprinting processes can be completed through different working mechanisms, such as extrusion-based cell-laden fiber deposition on a platform [82], coaxial nozzle-assisted crosslinking deposition, bioplotting in a plotting medium (i.e., crosslinker pool) [83], and precrosslinked alginate hydrogel coextruded with cells [84]. Each of these 3D bioprinting technologies has some merits in 3D printing of bioartificial organs.

In extrusion-based 3D bioprinting, great effort has been made to improve printing resolution and shape fidelity of the cell-laden alginate-containing 3D constructs by optimizing the processing parameters, such as nozzle size, dispensing pressure, and printing speed [85–89]. For example, Markstedt and coworkers blended alginate hydrogel with collagen and nanofibrillated cellulose to effectively enhance extrusion-based 3D printing resolution from 1000 to 400–600 μm using a proper diameter nozzle [90]. Kundu and coworkers printed cell-laden alginate with synthetic polycaprolactone (PCL) using a double-nozzle 3D bioprinter [14].

3.2. Gelatin

Gelatin is a partly hydrolyzed collagen derived from different animal tissues, such as bovine tendon and fish. Unlike its ancestor collagen, gelatin is a typical thermal-responsive (or thermosensitive) linear natural polymer with excellent water solubility, biocompatibility, biodegradability, and 3D printability. For example, the immunologic rejection of gelatin is much lower (i.e., zero) than that of collagen. There are no inflammation and other negative reactions when gelatin hydrogels are implanted in vivo [91]. This is an outstanding advantage of gelatin hydrogels for use as bioinks for 3D bioprinting of bioartificial organs.

The liquefaction temperature (i.e., sol–gel transition point) of gelatin solution is approximately 28 °C (25–30 °C). The unique sol–gel transition property and super biocompatibility of gelatin hydrogels have made them the preferential natural bioinks for 3D organ bioprinting. Before 3D bioprinting, cells and bioactive agents are homogeneously encapsulated in the gelatin solution to form cell-laden hydrogels. Other natural polymers, such as alginate, fibrinogen, chitosan, and hyaluronate, can be incorporated into the gelatin solution as additives to form composite bioinks [92–97]. Different chemical crosslinking strategies have been employed to improve the stability of 3D printed constructs according to the properties of the additives.

During the sol–gel transition stage, physical crosslinking occurs among the gelatin molecules. This means that thermosensitive gelatin solutions are capable of gelling and being 3D printed at mild or room temperatures, such as 1–28 °C, in which cells are durable. However, physical crosslinking (or gelling) is a reversible thermosensitive gelation process. The bonding strength among gelatin molecules is poor, which results in breakage of the 3D printed constructs when they are put into culture medium at physiological conditions, such as 37 °C. Some groups have utilized this phenomenon to print monolayer cells using extrusion-based 3D printers and washed the liquefied gelatin molecules post printing [98]. Otherwise, secondary chemical crosslinking is necessary to stabilize the 3D constructs [79–81].

Since 2005, various gelatin-based composite bioinks, such as gelatin/alginate, gelatin/fibrin, gelatin/chitosan, gelatin/hyaluronate, gelatin/alginate/fibrin, and gelatin/alginate/dextron (glaycerol or dimethyl sulfoxide), have been explored in my laboratory through various extrusion-based 3D bioprinting models (Figure 7) [15,32,79–81]. In these models, the viscosity of the gelatin-based bioinks depends largely on the polymer concentration, molecular weight, and cell density. A series of two-step stabilization strategies, containing both thermosensitive physical and ionic chemical crosslinks, have been exploited for 3D printed constructs. The chemical crosslinking methods include glutaraldehyde for gelatin, $CaCl_2$ for alginate, sodium tripolyphosphate for chitosan, and thrombin for fibrinogen.

Ten years on, these classical bioink formulations and crosslinking strategies have been widely adapted by many other groups all over the world [96,97].

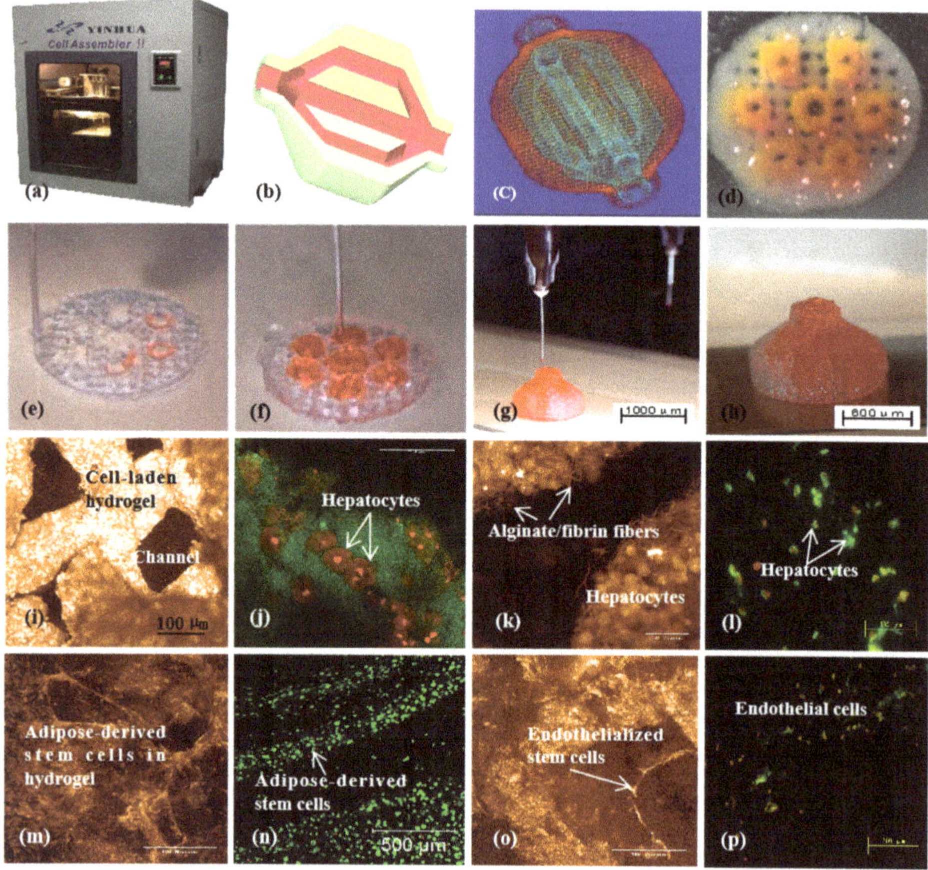

Figure 7. A large-scale 3D printed bioartificial organ with vascularized liver tissue constructed through the double-nozzle 3D bioprinter created in Prof. Wang's laboratory at Tsinghua University: (**a**) the double-nozzle 3D bioprinter; (**b**) a computer-aided design (CAD) model containing a branched vascular network; (**c**) a CAD model containing the branched vascular network; (**d**) a few layers of the 3D bioprinted construct containing both ASCs encapsulated in a gelatin/alginate/fibrin hydrogel and hepatocytes encapsulated in a gelatin/alginate/chitosan hydrogel; (**e**–**h**) 3D printing process of a semielliptical construct containing both ASCs and hepatocytes encapsulated in different hydrogels; (**i**–**l**) hepatocytes encapsulated in the gelatin-based hydrogels after 3D bioprinting and different periods of in vitro cultures; (**m**–**p**) ASCs encapsulated in the gelatin-based hydrogels after 3D bioprinting and different periods of in vitro cultures as well as growth factor inductions. Images reproduced with permission from [15,32].

When cells are embedded in gelatin-based solutions with high polymer concentration, their bioactivities are restricted to some degree after the double physical and chemical crosslinking. Meanwhile, lower concentrations of polymers facilitate higher cell viability. Nevertheless, when the concentration of gelatin-based polymers is reduced, the mechanical strength of the 3D constructs drops sharply despite chemical crosslinking. An optimized polymer concentration is necessary for a typical 3D bioprinting technology to ensure favorable cellular activity and structural stability [79–81].

Because the components of our gelatin-based composite bioinks are similar to ECMs (i.e., proteoglycans or glycoproteins), 3D printed constructs can be engineered to possess similar water content and elastic modulus as those of soft organs with excellent in vitro cell and in vivo tissue biocompatibilities. This is a fundamental breakthrough in large-scale 3D organ bioprinting. Until now, the resulting 3D printed constructs have been extended rapidly to other biomedical areas, such as high-throughput drug screening, living tissue and organ cryopreservation, metabolism model establishment, and pathological mechanism analysis [79–81].

Besides physical and chemical crosslinking strategies, some researchers have aimed to maintain laser-based 3D printed gelatin structures by methacrylation of the gelatin molecules (i.e., gelatin methacrylamide or gelatin methacryloyl, GelMA) using irreversible ultraviolet (UV) photopolymerization techniques [99]. Some researchers have blended GelMA with poly(ethylene glycol) (PEG) and other polymers. It was reported that the extruded strut size could be reduced from 1100–1300 to 350–450 μm by blending GelMA with PEG and to 150–200 μm by coaxial extrusion of GelMA with alginate using a coaxial nozzle [100–103]. Additional concerns have arisen regarding the toxicity of polymerization initiators and degradation products of synthetic polymethacrylamide.

3.3. *Fibrin*

Fibrin is a blood-derived fibrous protein formed by fibrinogen polymerization under the catalysis of thrombin, calcium ions, and/or factor XIII. Fibrinogen powder can dissolve in water to form a solution with low viscosity. Like gelatin, fibrin has excellent biocompatibility and biodegradability for various biomedical applications, such as hemostasia, wound healing, pharmacy, microencapsulated cell delivery, tissue engineering, and 3D bioprinting [104].

As stated above, layer-by-layer bioprinting is a logical choice for the manufacture of stratified organs, such as the liver, heart, kidney, and lung, containing multiple cell types. 3D bioprinting of stratified organ replacements depends on bioinks with appropriate rheological and cytocompatible properties. An obvious shortcoming of the fibrinogen solution for 3D organ bioprinting is that its viscosity is too low to be piled up in layers even in very high concentration (e.g., 4% w/v) [55]. After polymerization, the 3D printed constructs undergo further contraction or shrinkage for homogeneous or heterogeneous tissue formation. Structural stability is the major concern for pure fibrinogen as a bioink for 3D organ bioprinting.

The first pertinent fibrin 3D bioprinting was reported in 2007, in which fibrinogen was used as an additive of gelatin-based bioinks [105,106]. By blending fibrinogen with thermoresponsive gelatin solutions, high-resolution 3D constructs with excellent cell viability (i.e., ≈ 100%) have been obtained. During the 3D printing process, gelatin molecules provide fast gelation and immediate postprinting structural fidelity, while fibrinogen molecules ensure long-term mechanical stability upon thrombin polymerization. Physical blending and chemical crosslinking of fibrinogen/gelatin solutions can prevent breakdown and shrinkage of 3D printed constructs for a relatively longer period under physiological conditions (e.g., 37 °C).

More than 10 years later, Hinton and colleagues printed a Ca^{2+}-mediated alginate/fibrinogen hydrogel into a gelatin slurry bath containing thrombin using a similar extrusion-based 3D printer and working principle [16,17]. When the alginate/fibinogen hydrogel was extruded from the 3D printer nozzle, it met the thrombin molecules in the bath and rapidly polymerized to solidify the Ca^{2+}-mediated alginate/fibrinogen hydrogel. This is another double crosslinking or gelling path for 3D bioprinting with fibrinogen-containing composite bioinks and is termed as "bioplotting" (Figure 3). With different gelling paths, cells in the fibrinogen-containing hydrogel can be assembled in layers according to the predesigned CAD models.

Similarly, Burmeister and colleagues delivered ASCs via PEG–fibrin, i.e., PEGylated fibrin (FPEG) hydrogel, as an adjunct to meshed split-thickness skin grafts in a porcine model [107]. They showed that ASCs delivered in FPEG could enlarge the blood vessel size in a dose-dependent way, wherein FPEG acts as both a porous scaffold to prevent contraction and an ASC-delivery vehicle to accelerate angiogenesis.

For 3D organ bioprinting, the printing properties of fibrinogen solutions can be easily tailored by varying the proportion of its component in the composite bioinks [108–111]. Thermosensitive polymers, such as gelatin and agar, support extrusion-based 3D bioprinting of a wide range of natural polymers, including fibrinogen, chitosan, alginate, and hyaluronate. Succeeding (or tandem) chemical crosslinking and polymerization of fibrinogen-containing composite bioinks facilitates the printing results of the stratified cell-laden 3D constructs with increased stiffness and stabilization. For example, double gelation of gelatin/alginate/fibrinogen bioinks using both $CaCl_2$ (for alginate ion crosslinking) and thrombin (for fibrinogen polymerization) is the most popular approach today for large-scale organ manufacturing (Figure 8) [30,36].

Figure 8. 3D bioprinting of ASC-laden gelatin/alginate/fibrin hydrogel for organ manufacturing in Prof. Wang's laboratory at Tsinghua University: (**a**) a pioneering double-nozzle 3D bioprinter made in this laboratory; (**b**) schematic description of the cell-laden gelatin/alginate/fibrin hydrogel and pancreatic islets being printed into a grid construct using the 3D bioprinter; (**c**) a large-scale 3D printed grid construct containing ASC-laden gelatin/alginate/fibrin hydrogel cultured in a plate; (**d**) a grid ASC-laden gelatin/alginate/fibrin construct after being cultured for one month; (**e**) a multicellular construct after three weeks of culture, containing both ASCs encapsulated in the gelatin/alginate/fibrin hydrogel before epidermal growth factor (EGF) engagement and relatively integrated pancreatic islets seeding in the predefined channels (immunostaining with anti-insulin in green); (**f**) some envelopes of the islets were broken after one month of culture; (**g**) immunostaining of the 3D construct with mAbs for $CD31^+$ cells (i.e., mature endothelial cells from the ASC differentiation after three days of culture with EGF added in the culture medium) in green, having a fully confluent layer of endothelial cells (i.e.,

endothelium) on the surface of the predefined channels; (**h**) a vertical image of the 3D construct showing the fully confluent endothelium (formed from endothelial cells) and the predefined go-through channels; (**i**) immunostaining of the 3D construct with mAbs for CD34$^+$ cells (i.e., endothelial cells) in green and pyrindine (PI) for cell nuclei (nucleus) in red; (**j**) immunostaining of the 3D construct with mAbs for CD34$^+$ endothelial cells in green and PI for cell nuclei (nucleus) in red after three days of culture without EGF added in the culture medium; (**k**) immunostaining of the 3D construct with mAbs for CD31$^+$ endothelial cells in green and PI for cell nuclei (nucleus) in red after three days of culture with EGF added in the culture medium; (**l**) a control of (k), immunostaining of the 3D construct with mAbs for CD31$^+$ endothelial cells differentiated from the ASCs in green and PI for cell nuclei (nucleus) in red after three days of culture without EGF added in the culture medium; (**m**) immunostaining of the 3D construct with mAbs for CD31$^+$ cells in green and Oil Red O staining for adipocytes in red, showing both the heterogeneous tissues coming from the ASC differentiation after a cocktail growth factor engagement (i.e., on the surface of the channels, the endothelium coming from the ASCs differentiation after being treated with EGF for 3 days; deep inside the gelatin/alginate/fibrin hydrogel, the adipose tissue coming from the ASC differentiation after being subsequently treated with insulin, dexamethasone, and isobutylmethylxanthine (IBMX) for another three days); (**n**) a control of (m), showing all the ASCs in the 3D construct differentiated into target adipose tissue after three days of treatment with insulin, dexamethasone, and IBMX but no EGF; (**o**) immunostaining of two-dimensional (2D) cultured ASCs with mAbs for CD31$^+$ endothelial cells in green and PI for cell nuclei (nucleus) in red after three days of culture with EGF added in the culture medium; (**p**) immunostaining of 2D cultured ASCs with mAbs for CD31$^+$ endothelial cells in green and PI for cell nuclei (nucleus) in red after three days of culture without EGF added in the culture medium. Images reproduced with permission from [36].

3.4. dECM

dECM is a mixture of natural polymers, such as collagen and glycosaminoglycans (GAGs), derived from decellularization of different animal tissues, such as the heart, skin, liver, and small intestinal submucosa. The decellularization process can be chemical, physical, enzymatic, or their combinations, and it plays an important role in the final dECM compositions and geometries. After decellularization, the ECM compositions and geometries of the original tissues can be highly preserved, providing specific micro/meso/macrobiophysical and biochemical environments for different cell lines. In particular, dECM is rich in cell growth factors and stem cell niches, which help to support special tissue generation. dECM has remarkable advantages in 3D bioprinting of customized tissues due to the preservation of personal ECMs.

Compared with gelatin and alginate, the viscosity of dECM solution is rather low. Like cell-laden bioinks, dECM-based solutions gel rapidly beyond 15 °C and form physically crosslinked hydrogels upon increase in the temperature and pH [112]. Like most other natural polymers, such as alginate, gelatin, and fibrinogen, dECM-based bioinks can form semipermeable substrates with entrapped viable cells and biomolecules for 3D organ bioprinting. Water, gases, nutrients, metabolites, and growth factors can infiltrate into the semipermeable substrates. Human ASC (hASC) viability can reach 90% on the 14th day in the 3D bioprinted dECM substrates [113].

During the 3D bioprinting process, dECM displays poor 3D printability, low accuracy (or resolution), and large shrinkage due to the low viscosity. Various attempts have been made to surmount the shortcomings. The attempts include chemical crosslinking and concomitantly printing dECM with other natural and/or synthetic polymers. For example, Pati and coworkers printed an adipose-derived dECM–PCL hybrid construct with ASCs encapsulated in the dECM using a two-head tissue building system. Cell viability in this hybrid construct was >90%, equal to that without PCL [114].

Though dECM has remarkable advantages for tissue-specific function preservation, it faces many challenges for complex 3D organ bioprinting due to the following reasons. Firstly, it is difficult to efficiently remove the immunogen existing in the allogeneic or xenogeneic dECMs to avoid immune responses of the host tissues. Secondly, more or less residual deoxyribonucleic acid (DNA) or nuclear materials retained in dECMs can affect the cellular behaviors of the adopted cells. Thirdly, the extremely weak mechanical property, poor shape fidelity, and rapid degradation rate are major issues for large vascularized and innervated organ construction [115–117].

4. Synthetic Polymers for 3D Organ Bioprinting

Synthetic polymers are human-made polymers produced by chemical reactions of monomers, which may be derived from petroleum oil. Their hydrogels are generally produced via bulk, solution, and inverse dispersion procedures. While the first two procedures are homogeneous, the inverse dispersion procedure is conducted in dispersed and continuous phases [118–122]. Among homogeneous polymerizations, the solution reaction is preferred due to better control of the heat of polymerization and hence the polymer properties. Most high-swelling synthetic polymeric hydrogels are produced in this way.

As a common characteristic, the glass transition temperatures of synthetic polymer solutions, such as poly(lactic-co-glycolic acid) (PLGA) and polyurethane (PU), are much lower than those of the abovementioned nature polymers. Temperatures of −20 to −40 °C are common for most biodegradable synthetic polymeric solutions to solidify. Meanwhile, very high temperatures, such as 100–200 °C, are necessary for biodegradable synthetic polymers to melt. For example, polylactic acid (PLA) has a melting temperature of 180 °C. Thus, synthetic polymers are comparatively bioinert and do not readily embody bioactive ingredients, such as cells and growth factors, directly for 3D bioprinting. This is mainly due to the fact that 3D printing processes of synthetic polymers often involve the use of organic solvents, heat, or poisonous crosslinkers, which may reduce the bioactivity of the ingredients [79–81].

Compared with natural polymers, synthetic polymers are prominent in their mechanical properties as their molecular weights can be regulated from low to ultrahigh according to the actual requirement. High molecular weights promote intermolecular interactions between the polymer chains and have better mechanical properties for in vitro pulsatile culture using peristaltic pump and in vivo implantation of the 3D printed constructs. Nevertheless, most synthetic polymer solutions, hydrogels, and scaffolds, such as PLGA, poly(glycolic acid) (PGA), poly(hydroxypropyl methacrylamide) (PHPMA), PU, PCL, PLA, and poly(methyl methacrylate) (PMMA), have poor cytocompatibilities due to the intrinsic bioinert characteristics, organic solvent usages, and stiff morphological/topological structures [123].

The lack of functional groups and structural complexity within synthetic polymers has limited their usage in 3D organ bioprinting. Till now, most synthetic polymers have been applied as supporting or sacrificing structures without directly contacting living cells. In this section, three synthetic polymers, namely, PEG, PLGA, and PU, with excellent 3D printability, in vivo tissue compatibility (or bioinertia), and structural stability for 3D organ bioprinting are introduced.

4.1. PEG

PEG, also named as polyoxyethylene or poly(ethylene oxide) (PEO), is a biocompatible, nonimmunogenic synthetic polyether that has been approved by the Food and Drug Administration (FDA) of the United States as a good candidate for cell encapsulation and other biomedical applications. It is a hydrophilic polymer with linear and branched structures. PEG can be crosslinked using physical, ionic, or covalent crosslinks. Two hydroxyl groups of PEG diol can be tailored into other functional groups (i.e., acrylate, thiol, and carboxyl) by physical, ionic, or covalent crosslinking, which makes PEG possess tunable mechanical properties for 3D bioprinting [124,125].

Unlike other synthetic polymers, PEG appears in solid state at room temperature with a molecular weight (Mw) about 1000 Da. PEG is water-soluble, and the viscosity depends on the Mw and on the amount of water. PEG itself cannot form hydrogel. The low viscosity of PEG solutions makes it impossible for use in extrusion-based 3D bioprinting. Acrylation or blending with other polymers is often necessary for the development of a PEG-containing hydrogel. For example, Gao and coworkers have extensively studied the properties of PEG–GelMA for inkjet-based bioprinting [126,127].

The resulting constructs could be used for hard tissue regeneration with a reasonable mechanical strength (e.g., compressive modulus: 1–2 MPa). GelMA in the bioink could promote mesenchymal stem cell differentiation into cartilage and bone tissues. This is due to the existence of natural gelatin component in the GelMA molecules. Similarly, the poly(ethylene glycol) diacrylate (PEG-DA) hydrogels photopolymerized by UV light could increase the mechanical properties and printing resolutions of 3D constructs but without enhanced biological functions [128].

Though PEG and its derivatives have been employed in 3D bioprinting, the lack of cell-adhesive domains and poor mechanical properties have seriously limited their application in 3D organ bioprinting. A great effort has been made to improve the biological and physiological functions of 3D printed constructs. One of the most widely used strategies is to incorporate bioactive molecules, especially arginyl–glycyl–aspartate (or Arg–Gly–Asp, RGD) peptide segments, in the bioinks to promote cell activities. For example, Villanueva and coworkers investigated the role of cell–matrix interactions by dynamically loading cells in a RGD-incorporated PEG 3D printing process. The attachment (e.g., adhesion) and differentiation capabilities of the cells, such as osteoblasts, mesenchymal stem cells, endothelial cells, and smooth muscle cells, on the 3D scaffold augmented in a dose-dependent manner. Through dynamic 3D bioprinting, the RGD-incorporated PEG bioink could obviously enhance the chondrocyte phenotype and ECM synthesis, indicating that cell–matrix interactions directly mediated cell activities [128]. Actually, RGD is a special peptide segment containing three amide linkages, which exists in all natural proteins, including collagen and its derivative gelatin. It has the same adhesive properties for cells as those of natural polymers.

4.2. PLGA

PLGA is a synthetic copolymer (i.e., linear aliphatic polyester) of lactic acid (α-hydroxy propanoic acid) and glycolic acid (hydroxy acetic acid), which has been approved by FDA for therapeutic devices. It is synthesized by means of ring-opening copolymerization of two different monomers: the cyclic dimers (1,4-dioxane-2,5-diones) of lactic acid and glycolic acid [129]. PLGA 75:25 has been identified as a commonly used copolymer with a composition of 75% lactic acid and 25% glycolic acid. The monomer lactic acid contains an asymmetric carbon atom and therefore has two optical isomers: l (+) lactic acid and d (61) lactic acid. It is widely distributed in all living creatures (such as animals, human bodies, plants, and microorganisms) as either an intermediate or an end product in carbohydrate metabolism. Meanwhile, glycolic acid occurs in nature to a limited extent.

PLGA can be hydrolyzed by breaking the ester linkages in its chains under the presence of water. The final degradation products of PLGA are either acidic monomers, such as lactic acid and glycolic acid, or innocuous salts, such as lactate (salt form of lactic acid) and glycolate (salt form of glycolic acid). It has been shown that the time required for the degradation of PLGA is related to the monomers' ratio, which can be reflected in the molecular composition. The higher the content of glycolide units, the lower the time required for the degradation compared to the lactide predominant polymers [130].

PLGA dissolves in a wide range of organic solvents depending on its composition. Higher lactide-containing polymers dissolve in chlorinated solvents, whereas higher glycolide-containing polymers require the use of fluorinated solvents, such as 1,1,1,3,3,3-hexafluoroisopropanol. PLGA solutions show a typical glass transition temperature in the range of −40 to −60 °C.

Within the author's own group, we have developed various low-temperature RP technologies to deposit synthetic PLGA solutions alone or with other polymers. Different material systems can be 3D printed together using double or multinozzle 3D bioprinters, resulting in hybrid constructs, such as PLGA–gelatin, PLGA–collagen, PLGA/hydroxyapatite–PLGA/hydroxyapatite/phosphralated chitosan, with strong mechanical properties, tunable biodegradabilities, and acceptable in vivo biocompatibilities (Figure 9) [20,50,131–133]. The 3D printed constructs have been extensively used for bone, cartilage, nerve, liver, and other large organ repair/regeneration/replacement/restoration. At the same time, the concept of vascularization and neuralization of large-scale 3D printed tissues has been adapted rapidly all over the world [134–137].

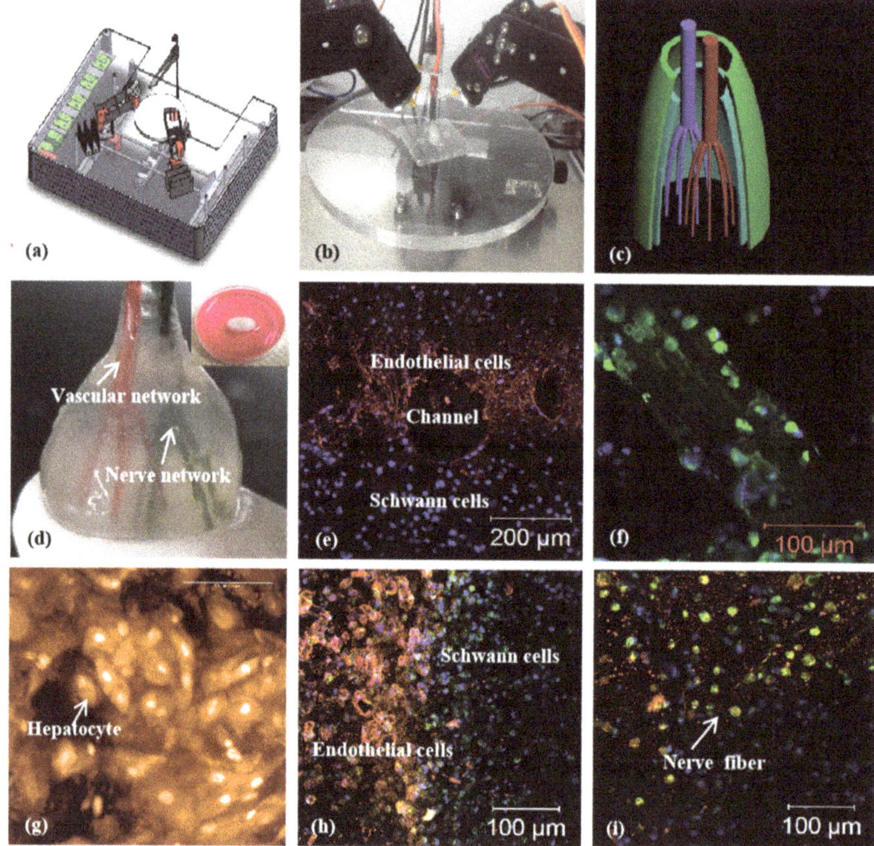

Figure 9. A combined four-nozzle 3D organ bioprinting technology created in Prof. Wang's laboratory at Tsinghua University in 2013 [20,50]: (**a**) equipment of the combined four-nozzle 3D organ bioprinter; (**b**) working state of the combined four-nozzle 3D organ printer; (**c**) a CAD model representing a large-scale vascularized and innervated hepatic tissue; (**d**) a semielliptical 3D construct containing a poly(lactic-co-glycolic acid) (PLGA) overcoat, a hepatic tissue made from hepatocytes in a gelatin/chitosan hydrogel, a branched vascular network with fully confluent endothelialized ASCs on the inner surface of the gelatin/alginate/fibrin hydrogel, and a hierarchical neural (or innervated) network made from Shwann cells in the gelatin/hyaluronate hydrogel; the maximal diameter of the semiellipse can be adjusted from 1 mm to 2 cm according to the CAD model; (**e**) a cross section of (d), showing the endothelialized ASCs and Schwann cells around a branched channel; (**f**) a large bundle of nerve fibers formed in (d); (**g**) hepatocytes underneath the PLGA overcoat; (**h**) an interface between the endothelialized ASCs and Schwann cells in (d); (**i**) some thin nerve fibers.

4.3. PU

PU is a family of synthetic polymers that are composed of organic units and joined by carbamate (i.e., urethane) links. PUs can be classified into two groups: biodegradable or nonbiodegradable. The traditional PUs are thermoresponsive (or thermoplastic, i.e., melt when heated) nonbiodegradable polymers that do not biodegrade when implanted in vivo. Historically, nonbiodegradable (or unbiodegradable) PU has been widely used in some biomedical fields due to its excellent mechanical and bioinert properties. Two examples of these biomedical applications are intravenous perfusion tubes and inanimate artificial hearts with strong mechanical strength [138,139]. However, these polymers cannot be printed using the existing 3D bioprinters.

In 2006, a brand new biodegradable elastomeric PU was developed by my group. This new PU is made of PEG and PCL monomers with excellent biocompatibility, biodegradability, bioprintability, and biostability for complex bioartificial organ manufacturing [136–138]. It can be 3D printed alone or with some other natural or synthetic polymers, such as gelatin, collagen, gelatin/alginate, etc. In one of our former studies, a hybrid hierarchical PU–cell/hydrogel construct was automatically created using an extrusion-based double-nozzle, low-temperature 3D printer (Figure 10) [28,49]. The PU mainly acted as a supportive template for cell, especially ASC, accommodation, growth, migration, proliferation, and differentiation. It is extremely useful for some antistress or antisuture applications, such as pulsatile culture of 3D printed hierarchical vascular and neural networks and in vivo anastomosis with the host vasculatures/neural networks [108–110]. The vascularized and innervated networks can be applied to 3D bioprinting of a variety of complex organs, such as the brain, heart, lung, and kidney.

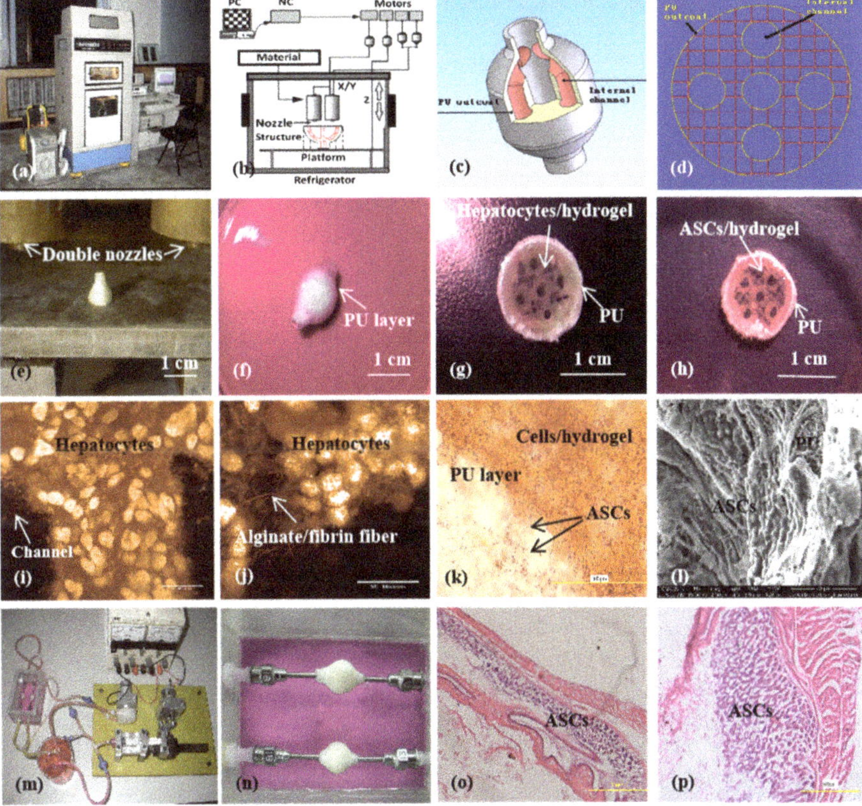

Figure 10. A large-scale 3D printed complex organ with vascularized liver tissue constructed through the double-nozzle low-temperature 3D bioprinter created in Prof. Wang's laboratory at Tsinghua University: (**a**) the double-nozzle low-temperature 3D bioprinter; (**b**) working principle of an elliptical hybrid hierarchical polyurethane-cell/hydrogel construct built via the double-nozzle low-temperature 3D bioprinter; (**c**) a CAD model containing the branched vascular network; (**d**) a cross section of the CAD model containing five sub-branched channels; (**e**) working platform of the 3D bioprinter containing two nozzles and the 3D printed semielliptical hybrid hierarchical polyurethane-cell/hydrogel

construct; (**f**) an elliptical sample containing both a cell-laden natural hydrogel and a synthetic polyurethane (PU) overcoat; (**g**) several layers of the elliptical sample in the middle section containing a hepatocyte-laden gelatin-based hydrogel and a PU overcoat; (**h**) several layers of the elliptical sample in the middle section containing an ASC-laden gelatin-based hydrogel and a PU overcoat; (**i**) hepatocytes encapsulated in the gelatin-based hydrogel; (**j**) a magnified photo of (i), showing the alginate/fibrin fibers around the hepatocytes; (**k**) ASCs encapsulated in the gelatin-based hydrogel growing into the micropores of the PU layer; (**l**) ASCs on the inner surface of the branched channels; (**m**) pulsatile culture of two elliptical samples; (**n**) two samples cultured in the bioreactor; (**o**) static culture of the ASCs encapsulated in the gelatin-based hydrogel; (**p**) pulsatile culture of the ASCs encapsulated in the gelatin-based hydrogel. Images reproduced with permission from [28,49].

Using similar techniques, a 3D printed PU/PEO scaffold was reported later in 2014. The PEO component could be used to increase the viscosity of the PU solution. The optimal ratio of PU/PEO for an extrusion-based FDM technology was 76/24. A resulting PU/PEO scaffold, containing an average pore size of 1–4 μm, was helpful to maintain cell morphology attached to it. Chondrocytes preferred to adhere onto the scaffold due to its hydrophilic property and suitable pore size [139]. In another study, neural stem cells (NSCs) attached well on the PU-based scaffold. The PU-based scaffold could provide NSCs with a proper environment to adhere, proliferate, and migrate on it and repair the disordered or damaged central nerve directly [140].

The control of multinozzle 3D printer parameters can provide a strong connection between different polymer systems, such as the PLGA solution–gelatin hydrogel, and PU solution–cell-laden gelatin-based hydrogel. Supportive synthetic polymers, such as PLGA and PU, and the ECM-like gelatin-based hydrogels can be adjusted to degrade at different time points during the organ construction and maturation stages. With proper selection of natural and synthetic polymer components, the 3D printed bioartificial organs can avoid all the risks of vascular rupture, stress shrinkage, immune rejection, and other negative reactions during the in vivo implantation stages [108–110]. These are all long-awaited breakthroughs in bioartificial organ manufacturing, as well as other pertinent hot research areas, such as tissue engineering, biomaterials, drug screening, organ transplantation, and pathological analysis, with respect to in vitro complex organ automatic building processes and in vivo defective/failed organ repair/regeneration/replacement/restoration applications.

5. Challenges and Perspectives

Despite remarkable progress achieved in 3D bioprinting over the last decade, there are still a number of challenges that remain in the manufacturing of off-the-shelf bioartificial organs for clinical applications. These challenges include (1) the extreme difficulties in establishment of an axial anastomosed vasculature, which anatomically mimics those in a natural organ, providing the incorporated cells with water, gas, and nutrients and removing metabolites from the cells; (2) the limited sophisticated multinozzle 3D printers, which are capable of assembling as many homogeneous and heterogeneous cells along with different polymers in a way similar to those in a natural organ; (3) the unmatched physical, biological, and physiological properties of 3D printed vascular, neural, and lymphatic networks for defective/failed organ restoration; (4) the uncertain roles of superfluous stem cells, growth factors, and other bioactive agents existing in the bioartificial organs for in vivo implantation [141–143].

In the future, the demand for 3D organ bioprinting, both for production in large amounts and small quantities as well as individual manufacture (i.e., customized), will definitely grow. More research needs to focus on mimicking the complex anatomical, material, and biological aspects of each human organ to be replaced, especially the hierarchical vascular, neural, and lymphatic networks for fluid and bioactive signal transportation. It is imperative to develop new sophisticated polymers with updating 3D bioprinters to recapitulate more complex micro-, meso-, and macro-3D milieus of natural organs with respect to geometrical features, material constituents, and environmental factors [144–146].

In the future, suitable choice of natural and synthetic polymers for 3D bioprinting of each bioartificial organ will still play a critical role. Much more integration of the existing natural and synthetic polymers for additional physiological functionality realization in a special natural organ is still on the way. High cell viability is a crucial prerequisite for complex 3D organ bioprinting with a large volume of homogeneous and heterogeneous cells. It is necessary to further clarify the special usages of various polymers for different tissue incorporation. Harsh 3D bioprinting conditions should be avoided or overcome through additional approaches. Patient-specific cells, especially stem cells and ECMs (or dECMs), are preferable to eliminate the immune reactions [147–152].

In the future, much more work needs to be done on stem cell extraction, culture, and induction. Because a human organ is composed of at least two types of cells, specific architectural design and multiple cell assembling are essential for each of the layer-by-layer 3D building processes. Sequential induction of stem cell differentiation in a predefined 3D construct needs to be normalized in a suitable laboratory [108–110]. In particular, ASCs have set an outstanding example for their vascularization and neuralization capabilities in 3D printed constructs. Further optimization of growth factors for different stem cell inductions needs to be carried out and standardized.

6. Concluding Remarks

3D organ bioprinting is a new, high-level interdisciplinary field that requires the integration of talents of many fields of science and technology, such as cell biology, computer, materials, information, chemistry, mechanics, engineering, manufacturing, and medicine. Bioartificial organs with tailored biological, biophysical, biochemical, and physiological properties can be 3D bioprinted through predesigned geometrical structures, biomaterial components, and processing parameters. Both natural and synthetic polymers have played essential and ubiquitous roles in successful 3D organ bioprinting technologies. Some natural polymer hydrogels, such as gelatin, fibrinogen, and dECM, have excellent cytocompatibilities due to their inorganic solvents, functional groups, and mild gelation conditions. Cells and bioactive agents can be incorporated in polymer solutions or hydrogels directly without significantly changing their viabilities and bioactivities. Biodegradable synthetic polymers, such as PLGA and PU, hold antisuture and antistress capabilities, which can be used as supporting structures for vascular and neural network strengthening in 3D printed constructs. The integration of natural and synthetic polymers using multinozzle 3D bioprinters in this author's own laboratory have paved all the rugged ways for bioartificial organ manufacturing with several series of successful 3D printing models. Further studies are still needed to develop more successful bioinks for 3D bioprinting of each human organ with a whole spectrum of physiological functions, such as multiple cell/ECM types, special geometrical shapes, and antisuture/antistress capabilities. Advanced polymer-based multidisciplinary efforts will reap much greater benefits in 3D organ bioprinting and will virtually replace failed/defective human organs in the near future.

Funding: The author acknowledges receipt of the following financial support for the research, authorship, and/or publication of this article: the National Natural Science Foundation of China (NSFC) (Nos. 81571832 and 81271665), the 2017 Discipline Promotion Project of China Medical University (CMU) (No. 3110117049), the Key Research and Development Project of Liaoning Province (No. 2018225082), and the 2018 Scientist Partners of China Medical University (CMU) and Shenyang Branch of Chinese Academy of Sciences (CAS) (No. HZHB2018013).

Conflicts of Interest: The author declares no competing financial interests.

References

1. Standring, S. (Ed.) *Gray's Anatomy*, 40th ed.; Churchill Livingstone: London, UK, 2008; ISBN 978-0-8089-2371-8.
2. Colunga, T.; Dalton, S. Building blood vessels with vascular progenitor cells. *Trends Mol. Med.* **2018**, *24*, 630–641. [CrossRef] [PubMed]
3. Wang, X.; Liu, C. 3D bioprinting of adipose-derived stem cells for organ manufacturing. In *Enabling Cutting Edge Technology for Regenerative Medicine*; Springer: Singapore, 2018; Chapter 1; pp. 3–14.

4. Aikawa, E.; Nahrendorf, M.; Sosnovik, D.; Lok, V.M.; Jaffer, F.A.; Aikawa, M.; Weissleder, R. Multimodality molecular imaging identifies proteolytic and osteogenic activities in early aortic valve disease. *Circulation* **2007**, *115*, 377–386. [CrossRef] [PubMed]
5. Moniaux, N.; Faivre, J.A. Rreengineered liver for transplantation. *J. Hepatol.* **2011**, *54*, 386–387. [CrossRef] [PubMed]
6. Orive, G.; Hernández, R.M.; Gascón, A.R. History, challenges and perspectives of cell microencapsulation. *Trends Biotechnol.* **2004**, *22*, 87–92. [CrossRef]
7. Wang, X.; Yan, Y.; Zhang, R. Rapid prototyping as tool for manufacturing bioartificial livers. *Trends Biotechnol.* **2007**, *25*, 505–513. [CrossRef]
8. Lei, M.; Wang, X. Biodegradable polymers and stem cells for bioprinting. *Molecules* **2016**, *21*, 539. [CrossRef]
9. Wang, X.; Ao, Q.; Tian, X.; Fan, J.; Wei, Y.; Hou, W.; Tong, H.; Bai, S. 3D bioprinting technologies for hard tissue and organ engineering. *Materials* **2016**, *9*, 802. [CrossRef]
10. Pourchet, L.J.; Thepot, A.; Albouy, M.; Courtial, E.J.; Boher, A.; Blum, L.J.; Marquette, C.A. Human skin 3D bioprinting using scaffold-free approach. *Adv. Healthcare Mater.* **2017**, *6*, 1601101. [CrossRef]
11. Norotte, C.; Marga, F.S.; Niklason, L.E.; Forgacs, G. Scaffold-free vascular tissue engineering using bioprinting. *Biomaterials* **2009**, *30*, 5910–5917. [CrossRef]
12. Rider, P.; Zhang, Y.; Tse, C.; Zhang, Y.; Jayawardane, D.; Stringer, J.; Callaghan, J.; Brook, I.M.; Miller, C.A.; Zhao, X.; et al. Biocompatible silk fibroin scaffold prepared by reactive inkjet printing. *J. Mater. Sci.* **2016**, *51*, 8625–8630. [CrossRef]
13. Christensen, K.; Xu, C.; Chai, W.; Zhang, Z.; Fu, J.; Huang, Y. Freeform inkjet printing of cellular structures with bifurcations. *Biotechnol. Bioeng.* **2015**, *112*, 1047–1055. [CrossRef] [PubMed]
14. Kundu, J.; Shim, J.-H.; Jang, J.; Kim, S.-W.; Cho, D.-W. An additive manufacturing-based PCL–alginate–chondrocyte bioprinted scaffold for cartilage tissue engineering. *J. Tissue Eng. Regen. Med.* **2015**, *9*, 1286–1297. [CrossRef] [PubMed]
15. Li, S.; Xiong, Z.; Wang, X.; Yan, Y.; Liu, H.; Zhang, R. Direct fabrication of a hybrid cell/hydrogel construct by a double-nozzle assembling technology. *J. Bioact. Compat. Polym.* **2009**, *24*, 249–265.
16. Noor, N.; Shapira, A.; Edri, R.; Gal, I.; Wertheim, L.; Dvir, T. 3D printing of personalized thick and perfusable cardiac patches and hearts. *Adv. Sci.* **2019**, *6*, 1900344. [CrossRef]
17. Grigoryan, B.; Paulsen, S.J.; Corbett, D.C.; Sazer, D.W.; Fortin, C.L.; Zaita, A.J.; Greenfield, P.T.; Calafat, N.J.; Gounley, J.P.; Ta, A.H.; et al. Multivascular networks and functional intravascular topologies within biocompatible hydrogels. *Science* **2019**, *364*, 458–464. [CrossRef] [PubMed]
18. Guillotin, B.; Souquet, A.; Catros, S.; Duocastella, M.; Pippenger, B.; Bellance, S.; Bareille, R.; Rémy, M.; Bordenave, L.; Amédée, J.; et al. Laser assisted bioprinting of engineered tissue with high cell density and microscale organization. *Biomaterials* **2010**, *31*, 7250–7256. [CrossRef]
19. Mannoor, M.S.; Jiang, Z.; James, T.; Kong, Y.L.; Malatesta, K.A.; Soboyejo, W.O.; Verma, N.; Gracias, D.H.; McAlpine, M.C. 3D Printed Bionic Ears. *Nano Lett.* **2013**, *13*, 2634–2639. [CrossRef]
20. Wang, J. Development of a Combined 3D Printer and Its Application in Complex Organ Construction. Master's Thesis, Tsinghua University, Beijing, China, 2014.
21. Wang, X.; Yan, Y.; Pan, Y.; Xiong, Z.; Liu, H.; Cheng, J.; Liu, F.; Lin, F.; Wu, R.; Zhang, R.; et al. Generation of three-dimensional hepatocyte/gelatin structures with rapid prototyping system. *Tissue Eng.* **2006**, *12*, 83–90. [CrossRef]
22. Yan, Y.; Wang, X.; Pan, Y.; Liu, H.; Cheng, J.; Xiong, Z.; Lin, F.; Wu, R.; Zhang, R.; Lu, Q. Fabrication of viable tissue-engineered constructs with 3D cell-assembly technique. *Biomaterials* **2005**, *26*, 5864–5871. [CrossRef]
23. Lei, M.; Wang, X. Uterus bioprinting. In *Organ Manufacturing*; Wang, X., Ed.; Nova Science Publishers Inc.: Hauppauge, NY, USA, 2015; pp. 335–355.
24. Li, S.; Yan, Y.; Xiong, Z.; Weng, C.; Zhang, R.; Wang, X. Gradient hydrogel construct based on an improved cell assembling system. *J. Bioact. Compat. Polym.* **2009**, *24*, 84–99. [CrossRef]
25. Zhao, X.; Liu, L.; Wang, J.; Xu, Y.F.; Zhang, W.M.; Khang, G.; Wang, X. In vitro vascularization of a combined system based on a 3D bioprinting technique. *J. Tissue Eng. Regen. Med.* **2014**, *10*, 833–842. [CrossRef] [PubMed]
26. Wang, J. Vascularization and adipogenesis of a spindle hierarchical adipose-derived stem cell/collagen/alginate-PLGA construct for breast manufacturing. *IJITEE* **2015**, *4*, 1–8.

27. Wang, X.; Huang, Y.W.; Liu, C. A combined rotational mold for manufacturing a functional liver system. *J. Bioact. Compat. Polym.* **2015**, *39*, 436–451. [CrossRef]
28. Huang, Y.; He, K.; Wang, X. Rapid Prototyping of a hybrid hierarchical polyurethane-cell/hydrogel construct for regenerative medicine. *Mater. Sci. Eng. C* **2013**, *33*, 3220–3229. [CrossRef]
29. Wang, X.; Sui, S.; Yan, Y.; Zhang, R. Design and fabrication of PLGA sandwiched cell/fibrin constructs for complex organ regeneration. *J. Bioact. Compat. Polym.* **2010**, *25*, 229–240. [CrossRef]
30. Yao, R.; Zhang, R.; Yan, Y.; Wang, X. In vitro angiogenesis of 3D tissue engineered adipose tissue. *J. Bioact. Compat. Polym.* **2009**, *24*, 5–24.
31. Xu, M.; Yan, Y.; Liu, H.; Yao, Y.; Wang, X. Control adipose-derived stromal cells differentiation into adipose and endothelial cells in a 3-D structure established by cell-assembly technique. *J. Bioact. Compat. Polym.* **2009**, *24*, 31–47. [CrossRef]
32. Liu, F.; Chen, Q.; Liu, C.; Ao, Q.; Tian, X.; Fan, J.; Tong, H.; Wang, X. Natural polymers for organ 3D bioprinting. *Ploymers* **2018**, *10*, 1278. [CrossRef]
33. Sui, S.; Wang, X.; Liu, P.; Yan, Y.; Zhang, R. Cryopreservation of cells in 3D constructs based on controlled cell assembly processes. *J. Bioact. Compat. Polym.* **2009**, *24*, 473–487. [CrossRef]
34. Wang, X.; Paloheimo, K.-S.; Xu, H.; Liu, C. Cryopreservation of cell/hydrogel constructs based on a new cell-assembling technique. *J. Bioact. Compat. Polym.* **2010**, *25*, 634–653. [CrossRef]
35. Wang, X.; Xu, H. Incorporation of DMSO and dextran-40 into a gelatin/alginate hydrogel for controlled assembled cell cryopreservation. *Cryobiology* **2010**, *61*, 345–351. [CrossRef] [PubMed]
36. Xu, M.; Wang, X.; Yan, Y.; Yao, R.; Ge, Y. A cell-assembly derived physiological 3D model of the metabolic syndrome, based on adipose-derived stromal cells and a gelatin/alginate/fibrinogen matrix. *Biomaterials* **2010**, *31*, 3868–3877. [CrossRef] [PubMed]
37. Wang, X.; Tuomi, J.; Mäkitie, A.A.; Poloheimo, K.-S.; Partanen, J.; Yliperttula, M. The integrations of biomaterials and rapid prototyping techniques for intelligent manufacturing of complex organs. In *Advances in Biomaterials Science and Applications in Biomedicine*; Lazinica, R., Ed.; InTech: Rijeka, Croatia, 2013; pp. 437–463.
38. Zhao, X.; Du, S.; Chai, L.; Xu, Y.; Liu, L.; Zhou, X.; Wang, J.; Zhang, W.; Liu, C.-H.; Wang, X. Anti-cancer drug screening based on an adipose-derived stem cell/hepatocyte 3D printing technique. *J. Stem Cell Res. Ther.* **2015**, *5*, 273.
39. Zhou, X.; Liu, C.; Zhao, X.; Wang, X. A 3D bioprinting liver tumor model for drug screening. *World J. Pharm. Pharm. Sci.* **2016**, *5*, 196–213.
40. Wang, X. Editorial: Drug delivery design for regenerative medicine. *Curr. Pharm. Des.* **2015**, *21*, 1503–1505. [CrossRef]
41. Wang, X. Overview on biocompatibilities of implantable biomaterials. In *Advances in Biomaterials Science and Biomedical Applications in Biomedicine*; Lazinica, R., Ed.; InTech: Rijeka, Croatia, 2013; pp. 111–155.
42. Wang, X.; Ma, J.; Wang, Y.; He, B. Bone repair in radii and tibias of rabbits with phosphorylated chitosan reinforced calcium phosphate cements. *Biomaterials* **2002**, *23*, 4167–4176. [CrossRef]
43. Wang, X.; Ma, J.; Feng, Q.; Cui, F. Skeletal repair in of rabbits with calcium phosphate cements incorporated phosphorylated chitin reinforced. *Biomaterials* **2002**, *23*, 4591–4600. [CrossRef]
44. Causa, F.; Sarracino, F.; De Santis, R. Basic structural parameters for the design of composite structures as ligament aumentation devices. *J. Appl. Biomater. Biomech.* **2006**, *4*, 21–30.
45. Fedorovich, N.E.; Schuurman, W.; Wijnberg, H.M.; Prins, H.-J.; van Weeren, P.R.; Malda, J.; Alblas, J.; Dhert, W.J.A. Biofabrication of osteochondral tissue equivalents by printing topologically defined, cell-laden hydrogel scaffolds. *Tissue Eng. Part C Methods* **2011**, *18*, 33–44. [CrossRef]
46. Hou, R.; Nie, L.; Du, G.; Xiong, X.; Fu, J. Natural polysaccharides promote chondrocyte adhesion and proliferation on magnetic nanoparticle/PVA composite hydrogels. *Colloids Surf. B Biointerfaces* **2015**, *132*, 146–154. [CrossRef]
47. Kong, H.J.; Kaigler, D.; Kim, K. Controlling rigidity and degradation of alginate hydrogels via molecular weight distribution. *Biomacromolecules* **2004**, *5*, 1720–1727. [CrossRef] [PubMed]
48. Wang, X.; Yan, Y.; Lin, F.; Xiong, Z.; Wu, R.; Zhang, R.; Lu, Q. Preparation and characterization of a collagen/chitosan/heparin matrix for an implantable bioartificial liver. *J. Biomater. Sci. Polym. E* **2005**, *16*, 1063–1080. [CrossRef] [PubMed]

49. Wang, X.; Ao, Q.; Tian, X.; Fan, J.; Wei, Y.; Tong, H.; Hou, W.; Bai, S. Gelatin-based hydrogels for organ 3D bioprinting. *Polymers* **2017**, *9*, 401. [CrossRef] [PubMed]
50. Li, S.; Tian, X.; Fan, J.; Tong, H.; Ao, Q.; Wang, X. Chitosans for tissue repair and organ three-dimensional (3D) bioprinting. *Micromachines* **2019**, *10*, 765. [CrossRef]
51. Lee, D.-Y.; Lee, H.; Kim, Y. Phage as versatile nanoink for printing 3-D cell-laden scaffolds. *Acta Biomater.* **2016**, *29*, 112–124. [CrossRef]
52. Hendriks, J.; Willem, V.C.; Henke, S.; Leijten, J.; Saris, D.B.; Sun, C.; Lohse, D.; Karperien, M. Optimizing cell viability in droplet-based cell deposition. *Sci. Rep.* **2015**, *11*, 11304. [CrossRef]
53. Benam, K.H.; Dauth, S.; Hassell, B.; Herland, A.; Jain, A.; Jang, K.-J.; Karalis, K.; Kim, H.J.; MacQueen, L.; Mahmoodian, R. Engineered in vitro disease models. *Annu. Rev. Pathol. Mech. Dis.* **2015**, *10*, 195–262. [CrossRef]
54. Gaetani, R.; Doevendans, P.A.; Metz, C.H.; Alblas, J.; Messina, E.; Giacomello, A.; Sluijter, J.P. Cardiac tissue engineering using tissue printing technology and human cardiac progenitor cells. *Biomaterials* **2012**, *33*, 1782–1790. [CrossRef]
55. Labbaf, S.; Ghanbar, H.; Stride, E.; Edirisinghe, M. Preparation of multilayered polymeric structures using a novel four-needle coaxial electrohydrodynamic device. *Macromol. Rapid Commun.* **2014**, *35*, 618–623. [CrossRef]
56. Melchels, F.P.; Feijen, J.; Grijpma, D.W. A review on stereolithography and its applications in biomedical engineering. *Biomaterials* **2010**, *31*, 6121–6130. [CrossRef]
57. Singh, S.; Afara, I.O.; Tehrani, A.H.; Oloyede, A. Effect of decellularization on the load-bearing characteristics of articular cartilage matrix. *Tissue Eng. Regen. Med.* **2015**, *12*, 294–305. [CrossRef]
58. Ye, L.; Zimmermann, W.-H.; Garry, D.J.; Zhang, J. Patching the heart cardiac repair from within and outside. *Circ. Res.* **2013**, *113*, 922–932. [CrossRef] [PubMed]
59. Fan, R.; Piou, M.; Darling, E.; Cormier, D.; Sun, J.; Wan, J. Bio-printing cell-laden matrigel-agarose constructs. *J. Biomater. Appl.* **2016**, *31*, 684–692. [CrossRef] [PubMed]
60. Panwar, A.; Tan, L.P. Current status of bioinks for micro-extrusion-based 3D bioprinting. *Molecules* **2016**, *21*, 685. [CrossRef]
61. Duan, B.; Kapetanovic, E.; Hockaday, L.A.; Butcher, J.T. Three-dimensional printed trileaflet valve conduits using biological hydrogels and human valve interstitial cells. *Acta Biomater.* **2014**, *10*, 1836–1846. [CrossRef]
62. Blaeser, A.; Duarte Campos, D.F.; Puster, U.; Richtering, W.; Stevens, M.M.; Fischer, H. Controlling shear stress in 3D bioprinting is a key factor to balance printing resolution and stem cell integrity. *Adv. Healthc. Mater.* **2016**, *5*, 326–333. [CrossRef]
63. Shim, J.H.; Lee, J.S.; Kim, J.Y.; Cho, D.W. Bioprinting of a mechanically enhanced three-dimensional dual cell-laden construct for osteochondral tissue engineering using a multi-head tissue/organ building system. *J. Micromech. Microeng.* **2012**, *22*, 085014. [CrossRef]
64. Yu, Y.; Zhang, Y.; Martin, J.A.; Ozbolat, I.T. Evaluation of cell viability and functionality in vessel-like bioprintable cell-laden tubular channels. *J. Biomech. Eng.* **2013**, *135*, 91011. [CrossRef]
65. Duan, B.; Hockaday, L.A.; Kang, K.H.; Butcher, J.T. 3D bioprinting of heterogeneous aortic valve conduits with alginate/gelatin hydrogels. *J. Biomed. Mater. Res. Part A* **2013**, *101A*, 1255–1264. [CrossRef]
66. Chung, J.Y.; Naficy, S.; Yue, Z.; Kapsa, R.; Quigley, A.; Moulton, S.E.; Wallace, G. Bio-ink properties and printability for extrusion printing living cells. *Biomater. Sci.* **2013**, *1*, 763–773. [CrossRef]
67. Liu, J.; Chi, J.; Wang, K.; Liu, X.; Gu, F. Full-thickness wound healing using 3D bioprinted gelatin-alginate scaffolds in mice: A histopathological study. *Int. J. Clin. Exp. Pathol.* **2016**, *9*, 11197–11205.
68. Luo, Y.X.; Luo, G.L.; Gelinsky, M.; Huang, P.; Ruan, C.S. 3D bioprinting scaffold using alginate/polyvinyl alcohol bioinks. *Mater. Lett.* **2017**, *189*, 295–298. [CrossRef]
69. Bendtsen, S.T.; Quinnell, S.P.; Wei, M. Development of a novel alginate-polyvinyl alcohol-hydroxyapatite hydrogel for 3D bioprinting bone tissue engineered scaffolds. *J. Biomed. Mater. Res. Part A* **2017**, *105*, 1457–1468. [CrossRef] [PubMed]
70. Yu, H.Y.; Ma, D.D.; Wu, B.L. Gelatin/alginate hydrogel scaffolds prepared by 3D bioprinting promotes cell adhesion and proliferation of human dental pulp cells in vitro. *J. South. Med. Univ.* **2017**, *37*, 668.
71. Nguyen, D.; Hägg, D.A.; Forsman, A.; Ekholm, J.; Nimkingratana, P.; Brantsing, C.; Kalogeropoulos, T.; Zaunz, S.; Concaro, S.; Brittberg, M.; et al. Cartilage tissue engineering by the 3D bioprinting of iPS cells in a nanocellulose/alginate bioink. *Sci. Rep.* **2017**, *7*, 658. [CrossRef]

72. Müller, M.; Öztürk, E.; Arlov, Ø.; Gatenholm, P.; Zenobi-Wong, M. Alginate sulfate-nanocellulose bioinks for cartilage bioprinting applications. *Ann. Biomed. Eng.* **2017**, *45*, 210–223. [CrossRef]
73. Jia, J.; Richards, D.J.; Pollard, S.; Tan, Y.; Rodriguez, J.; Visconti, R.P.; Trusk, T.C.; Yost, M.J.; Yao, H.; Markwald, R.R. Engineering alginate as bioink for bioprinting. *Acta Biomater.* **2014**, *10*, 4323–4331. [CrossRef]
74. Yu, J.; Du, K.T.; Fang, Q.; Gu, Y.; Mihardja, S.S.; Sievers, R.E.; Wu, J.C.; Lee, R.J. The use of human mesenchymal stem cells encapsulated in rgd modified alginate microspheres in the repair of myocardial infarction in the rat. *Biomaterials* **2010**, *31*, 7012–7020. [CrossRef]
75. Lee, K.Y.; Mooney, D.J. Alginate: Properties and biomedical applications. *Prog. Polym. Sci.* **2012**, *37*, 106–126. [CrossRef]
76. Pawar, S.N.; Edgar, K.J. Alginate derivatization: A review of chemistry, properties and applications. *Biomaterials* **2012**, *33*, 3279–3305. [CrossRef]
77. Tønnesen, H.H.; Karlsen, J. Alginate in drug delivery systems. *Drug Dev. Ind. Pharm.* **2002**, *28*, 621–630. [CrossRef] [PubMed]
78. Drury, J.L.; Dennis, R.G.; Mooney, D.J. The tensile properties of alginate hydrogels. *Biomaterials* **2004**, *25*, 3187–3199. [CrossRef] [PubMed]
79. Wang, X.; He, K.; Zhang, W. Optimizing the fabrication processes for manufacturing a hybrid hierarchical polyurethane-cell/hydrogel construct. *J. Bioact. Compat. Polym.* **2013**, *28*, 303–319. [CrossRef]
80. Liu, L.; Zhou, X.; Xu, Y.; Zhang, W.M.; Liu, C.-H.; Wang, X.H. Controlled release of growth factors for regenerative medicine. *Curr. Pharm. Des.* **2015**, *21*, 1627–1632. [CrossRef] [PubMed]
81. Xu, Y.; Wang, X. 3D biomimetic models for drug delivery and regenerative medicine. *Curr. Pharm. Des.* **2015**, *21*, 1618–1626. [CrossRef] [PubMed]
82. Axpe, E.; Oyen, M.L. Applications of alginate-based bioinks in 3D bioprinting. *Int. J. Mol. Sci.* **2016**, *17*, 1976. [CrossRef]
83. Park, J.; Lee, S.J.; Chung, S.; Lee, J.H.; Kim, W.D.; Lee, J.Y.; Park, S.A. Biofunctional rapid prototyping for tissue-engineering applications: 3D bioplotting versus 3D printing. *J. Polym. Sci. Part A Polym. Chem.* **2004**, *42*, 624–638.
84. Ahn, S.; Lee, H.; Bonassar, L.J.; Kim, G. Cells (MC3T3-E1)-laden alginate scaffolds fabricated by a modified solid-freeform fabrication process suppliemented with an aerosol spraying. *Biomacromolecules* **2012**, *13*, 2997–3003. [CrossRef]
85. Izadifar, Z.; Chang, T.; Kulyk, W.; Chen, X.; Eames, B.F. Analyzing biological performance of 3D-printed, cell-impregnated hybrid constructs for cartilage tissue engineering. *Tissue Eng. Part C Methods* **2016**, *22*, 173–188. [CrossRef]
86. Daly, A.C.; Cunniffe, G.M.; Sathy, B.N.; Jeon, O.; Alsberg, E.; Kelly, D.J. 3D bioprinting of developmentally inspired templates for whole bone organ engineering. *Adv. Healthc. Mater.* **2016**, *5*, 2353–2362. [CrossRef]
87. Jia, W.; Gungor-Ozkerim, P.S.; Zhang, Y.S.; Yue, K.; Zhu, K.; Liu, W.; Pi, Q.; Byambaa, B.; Dokmeci, M.R.; Shin, S.R.; et al. Direct 3D bioprinting of perfusable vascular constructs using a blend bioink. *Biomaterials* **2016**, *106*, 58–68. [CrossRef] [PubMed]
88. Murphy, S.V.; Atala, A. 3D bioprinting of tissues and organs. *Nat. Biotechnol.* **2014**, *32*, 773–785. [CrossRef] [PubMed]
89. Xu, W.; Wang, X.; Yan, Y.; Zhang, R. A polyurethane-gelatin hybrid construct for the manufacturing of implantable bioartificial livers. *J. Bioact. Compat. Polym.* **2008**, *23*, 409–422. [CrossRef]
90. Markstedt, K.; Mantas, A.; Tournier, I.; Martínez Ávila, H.; Hägg, D.; Gatenholm, P. 3D bioprinting human chondrocytes with nanocellulose–alginate bioink for cartilage tissue engineering applications. *Biomacromolecules* **2015**, *16*, 1489–1496. [CrossRef] [PubMed]
91. Wang, X.; Yu, X.; Yan, Y.; Zhang, R. Liver tissue responses to gelatin and gelatin/chitosan gels. *J. Biomed. Mater. Res. A* **2008**, *87A*, 62–68. [CrossRef] [PubMed]
92. Ng, W.L.; Yeong, W.Y.; Naing, M.W. Polyelectrolyte gelatin-chitosan hydrogel optimized for 3D bioprinting in skin tissue engineering. *Int. J. Bioprint.* **2016**, *2*, 1. [CrossRef]
93. Ribas, J.; Sadeghi, H.; Manbachi, A.; Leijten, J.; Brinegar, K.; Zhang, Y.S.; Ferreira, L.; Khademhosseini, A. Cardiovascular organ-on-a-chip platforms for drug discovery and development. *Appl. Vitro Toxicol.* **2016**, *2*, 82–96. [CrossRef]

94. Gaetani, R.; Feyen, D.A.; Verhage, V.; Slaats, R.; Messina, E.; Christman, K.L.; Giacomello, A.; Doevendans, P.A.; Sluijter, J.P. Epicardial application of cardiac progenitor cells in a 3D-printed gelatin/hyaluronic acid patch preserves cardiac function after myocardial infarction. *Biomaterials* **2015**, *61*, 339–348. [CrossRef]
95. Skardal, A.; Zhang, J.; McCoard, L.; Xu, X.; Oottamasathien, S.; Prestwich, G.D. Photocrosslinkable hyaluronan-gelatin hydrogels for two-step bioprinting. *Tissue Eng. Part A* **2010**, *16*, 2675–2685. [CrossRef]
96. Xiao, W.; He, J.; Nichol, J.W.; Wang, L.; Hutson, C.B.; Wang, B.; Du, Y.; Fan, H.; Khademhosseini, A. Synthesis and characterization of photocrosslinkable gelatin and silk fibroin interpenetrating polymer network hydrogels. *Acta Biomater.* **2011**, *7*, 2384–2393. [CrossRef]
97. Kang, H.-W.; Lee, S.J.; Ko, I.K.; Kengla, C.; Yoo, J.J.; Atala, A. A 3D bioprinting system to produce human-scale tissue constructs with structural integrity. *Nat. Biotechnol.* **2016**, *34*, 312–331. [CrossRef] [PubMed]
98. Lee, H.; Cho, D.-W. One-step fabrication of an organ-on-chip with spatial heterogeneity using a 3D bioprinting technology. *Lab Chip* **2016**, *16*, 2618–2625. [CrossRef] [PubMed]
99. Gauvin, R.; Chen, Y.-C.; Lee, J.W.; Soman, P.; Zorlutuna, P.; Nichol, J.W.; Bae, H.; Chen, S.; Khademhosseini, A. Microfabrication of complex porous tissue engineering scaffolds using 3D projection stereolithography. *Biomaterials* **2012**, *33*, 3824–3834. [CrossRef] [PubMed]
100. Mironov, V.; Kasyanov, V.; Drake, C.; Markwald, R.R. Organ printing: Promises and challenges. *Regen. Med.* **2008**, *3*, 93–103. [CrossRef]
101. Saunders, R.E.; Gough, J.E.; Derby, B. Delivery of human fibroblast cells by piezoelectric drop-on-demand inkjet printing. *Biomaterials* **2008**, *29*, 193–203. [CrossRef]
102. Phillippi, J.A.; Miller, E.; Weiss, L. Microenvironments engineered by inkjet bioprinting spatially direct adult stem cells toward muscle- and bone-like subpopulations. *Stem Cells* **2008**, *26*, 127–134. [CrossRef]
103. Kolessky, D.B.; Homan, K.A.; SkylarScott, M.A. Three-dimensional bioprinting of thick vascularized tissues. *PNAS* **2016**, *113*, 3179–3184. [CrossRef]
104. Mosesson, M.W. Fibrinogen and fibrin structure and functions. *J. Thromb. Haemost.* **2005**, *3*, 1894–1904. [CrossRef]
105. Xu, W.; Yan, Y.; Zheng, W.; Xiong, Z.; Lin, F.; Wu, R.; Zhang, R. Rapid prototyping of three-dimensional cell/gelatin/fibrinogen constructs for medical regeneration. *J. Bioact. Compat. Polym.* **2007**, *22*, 363–377. [CrossRef]
106. Liu, F.; Liu, C.; Chen, Q.; Ao, Q.; Tian, X.; Fan, J.; Tong, H.; Wang, X. Progress in organ 3D bioprinting. *Int. J. Bioprint.* **2017**, *4*, 1–15. [CrossRef]
107. Burmeister, D.M.; Stone, R.; Wrice, N.; Laborde, A.; Becerra, S.C.; Natesan, S.; Christy, R.J. Delivery of allogeneic adipose stem cells in polyethylene glycol-fibrin hydrogels as an adjunct to meshed autografts after sharp debridement of deep partial thickness burns. *Stem Cells Transl. Med.* **2018**, *7*, 360–372. [CrossRef] [PubMed]
108. Wang, X. Bioartificial organ manufacturing technologies. *Cell Transplant.* **2018**, *27*, 1–13. [CrossRef] [PubMed]
109. Wang, X.; Yan, Y.; Zhang, R. Recent trends and challenges in complex organ manufacturing. *Tissue Eng. Part B* **2010**, *16*, 189–197. [CrossRef] [PubMed]
110. Wang, X. Intelligent freeform manufacturing of complex organs. *Artif. Organs* **2012**, *36*, 951–961. [CrossRef]
111. Wang, X.; Liu, C. Fibrin hydrogels for endothelialized liver tissue engineering with a predesigned vascular network. *Polymers* **2018**, *10*, 1048. [CrossRef]
112. Sellaro, T.L.; Ranade, A.; Faulk, D.M.; McCabe, G.P.; Dorko, K.; Badylak, S.F.; Strom, S.C. Maintenance of human hepatocyte function in vitro by liver-derived extracellular matrix gels. *Tissue Eng. Part A.* **2010**, *16*, 1075–1082. [CrossRef]
113. Pati, F.; Ha, D.-H.; Jang, J.; Han, H.H.; Rhie, J.-W.; Cho, D.-W. Biomimetic 3D tissue printing for soft tissue regeneration. *Biomaterials* **2015**, *62*, 164–175. [CrossRef]
114. Pati, F.; Jang, J.; Ha, D.-H.; Kim, D.H.; Cho, D.W. Printing three-dimensional tissue analogues with decellularized extracellular matrix bioink. *Nat. Commun.* **2014**, *5*, 3935. [CrossRef]
115. Freytes, D.O.; Martin, J.; Velankar, S.S.; Lee, A.S.; Badylak, S.F. Preparation and rheological characterization of a gel form of the porcine urinary bladder matrix. *Biomaterials* **2008**, *29*, 1630–1637. [CrossRef]
116. Wong, S.Y.; Tay, C.Y.; Wen, F.; Venkatraman, S.S.; Tan, L.P. Engineered polymeric biomaterials for tissue engineering. *Curr. Tissue Eng.* **2012**, *1*, 41–53. [CrossRef]

117. Wolf, M.T.; Daly, K.A.; Brennan-Pierce, E.P.; Johnson, S.A.; Carruthers, C.; D'Amore, A.; Nagarkar, S.P.; Velankar, S.S.; Badylak, S.F. A hydrogel derived from decellularized dermal extracellular matrix. *Biomaterials* **2012**, *33*, 7028–7038. [CrossRef] [PubMed]
118. Vert, M.; Doi, Y.; Hellwich, K.-H.; Hess, M.; Hodge, P.; Kubisa, P.; Rinaudo, M.; Schué, F. Terminology for biorelated polymers and applications (IUPAC Recommendations 2012). *Pure Appl. Chem.* **2012**, *84*, 377–410. [CrossRef]
119. Vázquez-Rodriguez, G.A.; Beltránhernández, R.I.; Luchoconstantin, C.A.; Blasco, J.L. A Method for measuring the anoxic biodegradability under denitrifying conditions. *Chemosphere* **2008**, *71*, 1363–1368. [CrossRef] [PubMed]
120. Diaz-Santana, A.; Shan, M.; Stroock, A.D. Endothelial cell dynamics during anastomosis in vitro. *Integr. Biol.* **2015**, *7*, 454–466. [CrossRef]
121. Ng, H.Y.; Lee, K.-X.A.; Kuo, C.-N.; Shen, Y.-F. Biopring of artificial blood vessels. *Int. J. Bioprint.* **2018**, *4*, 140. [CrossRef]
122. Hoffman, A.S. Hydrogels for biomedical applications. *Adv. Drug Deliv. Rev.* **2002**, *43*, 3–10. [CrossRef]
123. Jungst, T.; Smolan, W.; Schacht, K.; Scheibel, T.; Groll, J.R. Strategies and molecular design criteria for 3D printable hydrogels. *Chem. Rev.* **2016**, *116*, 1496–1539. [CrossRef]
124. Zhu, J. Bioactive modification of poly (ethylene glycol) hydrogels for tissue engineering. *Biomaterials* **2010**, *31*, 4639–4656. [CrossRef]
125. Peppas, N.A.; Keys, K.B.; Torres-Lugo, M.; Lowman, A.M. Poly (ethylene glycol)-containing hydrogels in drug delivery. *J. Control. Release* **1999**, *62*, 81–87. [CrossRef]
126. Gao, G.; Schilling, A.F.; Hubbell, K.; Yonezawa, T.; Truong, D.; Hong, Y.; Dai, G.; Cui, X. Inkjet-bioprinted acrylated peptides and PEG hydrogel with human mesenchymal stem cells promote robust bone and cartilage formation with minimal printhead clogging. *Biotechnol. J.* **2015**, *10*, 1568–1577. [CrossRef]
127. Gao, G.; Schilling, A.F.; Hubbell, K. Improved properties of bone and cartilage tissue from 3D inkjet-bioprinted human mesenchymal stem cells by simultaneous deposition and photocrosslinking in PEG-GelMA. *Biotechnol. Lett.* **2015**, *37*, 2349–2355. [CrossRef]
128. Villanueva, I.; Weigel, C.A.; Bryant, S.J. Cell-matrix interactions and dynamic mechanical loading influence chondrocyte gene expression and bioactivity in PEG-RGD hydrogels. *Acta Biomater.* **2009**, *5*, 2832–2846. [CrossRef]
129. Astete, C.E.; Sabliov, C.M. Synthesis and characterization of PLGA nanoparticles. *J. Biomater. Sci. Polym. Ed.* **2006**, *17*, 247–289. [CrossRef] [PubMed]
130. Samadi, N.; Abbadessa, A.; Di Stefano, A.; van Nostrum, C.F.; Vermonden, T.; Rahimian, S.; Teunissen, E.A.; van Steenbergen, M.J.; Amidi, M.; Hennink, W.E. The effect of lauryl capping group on protein release and degradation of poly (D, L-lactic-co-glycolic acid) particle. *J. Contrl. Release* **2013**, *172*, 436–443. [CrossRef] [PubMed]
131. Liu, L.; Yan, Y.; Xiong, Z.; Zhang, R.; Wang, X. A novel poly (lactic-co-glycolic acid)-collagen hybrid scaffold fabricated via multi-nozzle low-temperature deposition. In *Virtual Rapid Manufacturing, Proceedings of the 3rd International Conference on Advanced Research in Virtual and Rapid Prototyping, Leiria, Portugal, 24–29 September 2007*; Bártolo, P., Ed.; © Taylor & Francis Group: London, UK, 2008; ISBN 978-0-415-41602-3.
132. Wang, X.; Rijff, B.L.; Khang, G. A building block approach into 3D printing a multi-channel organ regenerative scaffold. *J. Tissue Eng. Regen. Med.* **2015**, *11*, 1403. [CrossRef] [PubMed]
133. Wang, X. 3D printing of tissue/organ analogues for regenerative medicine. In *Handbook of Intelligent Scaffolds for Regenerative Medicine*, 2nd ed.; Pan Stanford Publishing: Palo Alto, CA, USA, 2016; pp. 557–570.
134. Marrella, A.; Lee, T.Y.; Lee, D.H.; Karuthedom, S.; Syla, D.; Chawla, A.; Khademhosseini, A.; Jang, H.L. Engineering vascularized and innervated bone biomaterials for improved skeletal tissue regeneration. *Mater. Today* **2018**, *21*, 362–376. [CrossRef]
135. Zhang, C.; Wen, X.; Vyavahare, N.R.; Boland, T. Synthesis and characterization of biodegradable elastomeric polyurethane scaffolds fabricated by the inkjet technique. *Biomaterials* **2008**, *29*, 3781–3791. [CrossRef]
136. Yan, Y.; Wang, X.; Yin, D.; Zhang, R.J. A new polyurethane/heparin vascular graft for small-caliber vein repair. *J. Bioact. Compat. Polym.* **2007**, *22*, 323–341. [CrossRef]
137. Yin, D.; Wang, X.; Yan, Y.; Zhang, R. Preliminary studies on peripheral nerve regeneration along a new polyurethane conduit. *J. Bioact. Compat. Polym.* **2007**, *22*, 143–159.

138. Xu, W.; Wang, X.; Yan, Y.; Zhang, R. Rapid prototyping of polyurethane for the creation of vascular systems. *J. Bioact. Compat. Polym.* **2008**, *23*, 103–114. [CrossRef]
139. Hung, K.C.; Tseng, C.S.; Hsu, S.H. Synthesis and 3D printing of biodegradable polyurethane elastomer by a water-based process for cartilage tissue engineering applications. *Adv. Healthc. Mater.* **2014**, *3*, 1578–1587. [CrossRef] [PubMed]
140. Hsieh, F.Y.; Hsu, S.H. 3D bioprinting: A new insight into the therapeutic strategy of neural tissue regeneration. *Organogenesis* **2015**, *11*, 153–158. [CrossRef] [PubMed]
141. Hinton, T.J.; Jallerat, Q.; Palchesko, R.N.; Park, J.H.; Grodzicki, M.S.; Shue, H.-J.; Ramadan, M.H.; Hudson, A.R.; Feinberg, A.W. Three-dimensinal printing of complex biological structures by freeform reversible embedding of suspended hydrogels. *Sci. Adv.* **2015**, *1*, e1500758. [CrossRef] [PubMed]
142. Xu, Y.; Wang, X. Liver manufacturing approaches: The thresholds of cell manipulation with bio-friendly materials for multifunctional organ regeneration. In *Organ Manufacturing*; Wang, X., Ed.; Nova Science Publishers Inc.: Hauppauge, NY, USA, 2015; pp. 201–225.
143. Wang, X. Spatial effects of stem cell engagement in 3D printing constructs. *J. Stem Cells Res. Rev. Rep.* **2014**, *1*, 5–9.
144. Shi, L.; Carstensen, H.; Hölzl, K.; Lunzer, M.; Li, H.; Hiborn, J.; Ovsianikov, A.; Ossipov, D.A. Dynamic coordination chemistry enables free directional printing of biopolymer hydrogel. *Chem. Mater.* **2017**, *29*, 5816–5823. [CrossRef]
145. Xu, Y.; Wang, X. Fluid and cell behaviors along a 3D printed alginate/gelatin/fibrin channel. *Bioeng. Biotech.* **2015**, *112*, 1683–1695. [CrossRef] [PubMed]
146. Liu, L.; Wang, X. Creation of a vascular system for complex organ manufacturing. *Int. J. Bioprint.* **2015**, *1*, 77–86. [CrossRef]
147. He, K.; Wang, X. Rapid prototyping of tubular polyurethane and cell/hydrogel constructs. *J. Bioact. Compat. Polym.* **2011**, *26*, 363–374.
148. Wang, X.; Cui, T.; Yan, Y.; Zhang, R. Peroneal nerve regeneration along a new polyurethane-collagen guide conduit. *J. Bioact. Compat. Polym.* **2009**, *24*, 109–127. [CrossRef]
149. Wang, X.; Sui, S. Pulsatile culture of a PLGA sandwiched cell/hydrogel construct fabricated using a step by step mold/extraction method. *Artif. Organs* **2011**, *35*, 645–655. [CrossRef]
150. Wang, X.; Yan, Y.; Zhang, R. Gelatin-based hydrogels for controlled cell assembly. In *Biomedical Applications of Hydrogels Handbook*; Ottenbrite, R.M., Ed.; Springer: New York, NY, USA, 2010; pp. 269–284.
151. Zhao, X.; Wang, X. Preparation of an adipose-derived stem cell/fibrin-poly (DL-lactic-co-glycolic acid) construct based on a rapid prototyping technique. *J. Bioact. Compat. Polym.* **2013**, *28*, 191–203. [CrossRef]
152. Cui, T.; Yan, Y.; Zhang, R.; Liu, L.; Xu, W.; Wang, X. Rapid prototyping of a double layer polyurethane-collagen conduit for peripheral nerve regeneration. *Tissue Eng. Part C* **2009**, *15*, 1–9. [CrossRef] [PubMed]

 *micromachines*

Review

Chitosans for Tissue Repair and Organ Three-Dimensional (3D) Bioprinting

Shenglong Li [1], Xiaohong Tian [1], Jun Fan [1], Hao Tong [1], Qiang Ao [1] and Xiaohong Wang [1,2,*]

[1] Center of 3D Printing & Organ Manufacturing, School of Fundamental Sciences, China Medical University (CMU), No. 77 Puhe Road, Shenyang North New Area, Shenyang 110122, China; lishenglong@cancerhosp-ln-cmu.com (S.L.); xhtian@cmu.edu.cn (X.T.); jfan@cmu.edu.cn (J.F.); tongh007@hotmail.com (H.T.); aoqiang00@163.com (Q.A.)

[2] Center of Organ Manufacturing, Department of Mechanical Engineering, Tsinghua University, Beijing 100084, China

* Correspondence: wangxiaohong709@163.com or wangxiaohong@tsinghua.edu.cn; Tel.: +86-24-3190-0983

Received: 24 October 2019; Accepted: 5 November 2019; Published: 11 November 2019

Abstract: Chitosan is a unique natural resourced polysaccharide derived from chitin with special biocompatibility, biodegradability, and antimicrobial activity. During the past three decades, chitosan has gradually become an excellent candidate for various biomedical applications with prominent characteristics. Chitosan molecules can be chemically modified, adapting to all kinds of cells in the body, and endowed with specific biochemical and physiological functions. In this review, the intrinsic/extrinsic properties of chitosan molecules in skin, bone, cartilage, liver tissue repair, and organ three-dimensional (3D) bioprinting have been outlined. Several successful models for large scale-up vascularized and innervated organ 3D bioprinting have been demonstrated. Challenges and perspectives in future complex organ 3D bioprinting areas have been analyzed.

Keywords: 3D bioprinting; chitosan; chitin; tissue repair; rapid prototyping (RP); implantable bioartificial organs

---

1. Introduction

An organ is a combination of multiple tissues that provide specific physiological functions for human body. There are about 80 organs in the human body, dominating every physiological activity [1]. Clinically, organ failure is the leading cause of mortality all over the world, and it is often closely related to some chronic and acute diseases [2,3]. As a result, one failing organ can break down the whole physiological system of the human body.

At present, allogeneic organ transplantation is the only cure for the failing internal organs, such as the liver, kidney, and heart. However, it is facing many limitations in clinical applications. First, there is a desperate shortage of donor organs. In another word, the donor organ is seriously short. Take the year of 2013 as an example, there were 117,040 patients in the United States of America who needed organ transplantation, but only 28,053 of them are fortunate to get suitable donors [4]. Currently, there are over 34 million surgical procedures in America involved in the treatment of organ failures per year. Less than one out of ten patients can be saved by organ donations [5]. Second, there are serious side effects of immunosuppressive drugs. The donor organs are allogeneic sources and the patients need to take a life-long immunosuppressive treatment. Since the side effects of immunosuppressive complications are severe, the ability of resisting infectious diseases can be obviously weakened. Third, the surgical cost is very high. Organ transplantation can bring huge economic burdens to the ordinary patients and occupy enormous medical resources [6].

During the last several decades, the severity of donor organ shortage, the life-long treatment of allograft rejection, the side effect of immunosuppressive therapy, and the extremely high cost of allogeneic organ transplantation have activated numerous strategies for tissue repair and organ manufacturing [7–11]. One typical example is tissue engineering. It has undergone several circulations of ups and downs with the three elements, i.e., porous scaffolds, cells, and growth factors. Nevertheless, organ manufacturing is a complex project that requires multi-disciplinary cooperation, involving a large scope of talents of technologies, such as biology, materials, chemistry, physics, mechanics, informatics, computers, and medicine. Designing and building the physical analogues of organs is only the first small step. It is more critical to make the multiple cell types/extracellular matrices (ECMs), hierarchical vascular, neural and/or lymphatic networks functional in a compacted construct [12–14].

Three-dimensional (3D) bioprinting has recently emerged as an extension of traditional rapid prototyping (RP), also named as solid freeform fabrication (SFF) and additive manufacturing (AM), technologies, by using bioactive or cellular components to build constructs in an additive or layer-by-layer methodology for encapsulation and culture of cells. These technologies allow for cell culture in controlled spatial environments. These environments can be tuned to simulate the complexity of in vivo cell growth environments with similar ECMs.

Polymers are large molecules or macromolecules, composed of many repeated subunits or small molecules, with molar masses ranging from thousands to millions. In another word, the units composing polymers derive from molecules of relatively low molecular mass [15]. There are two types of polymers: natural and synthetic. Natural polymers are naturally occurring polymers, such as cellulose, polysaccharide, protein, silk, and fibrinogen. Synthetic polymers are manmade polymers through chemically joining many small monomers together into one giant molecule. The list of synthetic polymers, roughly in order of worldwide demand, includes polyethylene, polypropylene, polystyrene, polyvinyl chloride, synthetic rubber, neoprene, nylon, polyacrylonitrile, phenol formaldehyde resin (or Bakelite), polyvinylbutyral (PVB), silicone, and many more. Because of their broad range of properties, both synthetic and natural polymers play essential and ubiquitous roles in everyday life. More than 330 million tons of these polymers are made every year (2015) [16]. Chitosan as a special natural polymer has attracted great attention both in tissue repair and organ 3D bioprinting areas.

2. Three-Dimensional (3D) Bioprinting

2.1. The Concept of Organ 3D Bioprinting

3D printing, traditionally termed as RP, SFF, and AM, is a series of material processing technologies based on the dispersion-accumulation principle of computer-aided manufacturing (CAM). Generally, an object can be divided into numerous two-dimensional (2D) layers before 3D printing with a defined thickness. The 2D layers are sequentially piled up by selectively adding the desired materials in a high reproductive additive manner under the instruction of computer-aided design (CAD) models [17–20].

Organ 3D bioprinting is the utilization of advanced 3D printing technologies to assemble multiple cell types, including stem cells/growth factors, along with other biomaterials in a layer-by-layer fashion to produce bioartificial organs that maximally imitate their natural counterparts with respect to anatomical structures, material components, and physiological functions (Figure 1) [12–14]. It mainly consists of four aspects, such as cell extraction (i.e., biological part), data collection (i.e., informatical part), starting material preparation (i.e., biomaterial part), and manufacturing (i.e., processing part). Patient-specific organ images, such as magnetic resonance imaging (MRI) and computerized tomography (CT) can be easily transferred into CAD models for customized organ manufacturing with predefined geometrical shapes, material (e.g., cells, growth factors, polymers, ECMs, drugs) components, and physiological functions [21–24]. Over the last decade, organ 3D bioprinting technologies have made a great contribution to various biomedical fields.

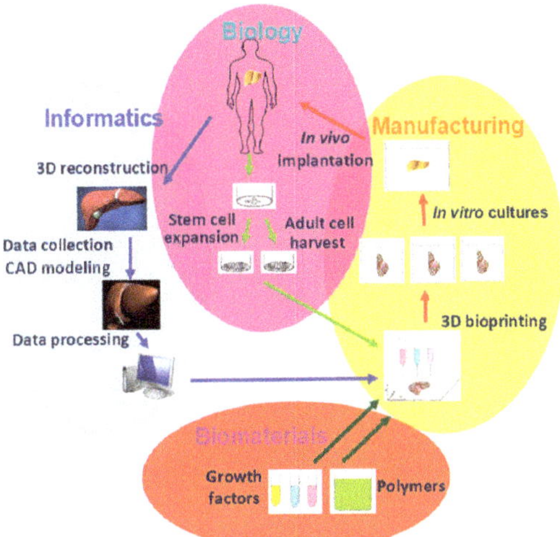

Figure 1. Graphical description of organ 3D bioprinting with four major aspects: cell extraction (i.e., biological), data collection (i.e., informatical), starting material preparation (i.e., biomaterial), and manufacturing (i.e., processing) parts.

2.2. Polymers as "Bioinks" for 3D Bioprinting

As stated above, polymers are large molecules made up of many small and identical repeating units bonded by covalent bonds. Theoretically, any polymer solutions or hydrogels holding the sol-gel transition property can be printed in layers under the instruction of CAD models [25,26]. Living cells and growth factors can be encapsulated into the polymeric solutions or hydrogels for 3D tissue and organ construction. Actually, only few natural and synthetic polymers and their combinations can be used as "bioinks" for tissue and organ 3D bioprinting at mild temperatures or cell endurable conditions. The layer-by-layer construction processes depend largely on the liquid polymer solution transformation capabilities before, during, and after the 3D bioprinting. There are many different sol-gel transformation forms, such as physical (reversible), chemical (reversible or irreversible), and biochemical (i.e., enzymic) cross-linking. Compared with synthetic polymers, most of the natural polymeric hydrogels can provide cells with suitable environments to survive. Currently, natural polymeric hydrogels are the dominate components of "bioinks" for 3D bioprinting.

There are three major types of organ 3D bioprinting technologies (Figure 2): (A) multi-nozzle extrusion-based bioprinting (a: pneumatic; b: Piston); (B) multi-nozzle inkjet-based bioprinting (a: heater; b: piezoelectric actuator); (C) multi-channel laser-assisted bioprinting. Cellular behaviors are easily manipulated within the polymeric hydrogels, via adjusting the physical, chemical, biochemical, and physiological properties of the 3D printable polymers. It is surprising that all the bottleneck problems, such as large scale-up tissue/organ manufacturing, living tissue/organ preservation, hierarchical vascular/neural network construction, and partly/fully stem cell engagement, which have perplexed tissue engineers and other researchers for more than several decades, have been overcome by a single scientist, the corresponding author of this article herself, via several series of automatic and semiautomatic layer-by-layer material integration and step-by-step stem cell inducement strategies [12–14].

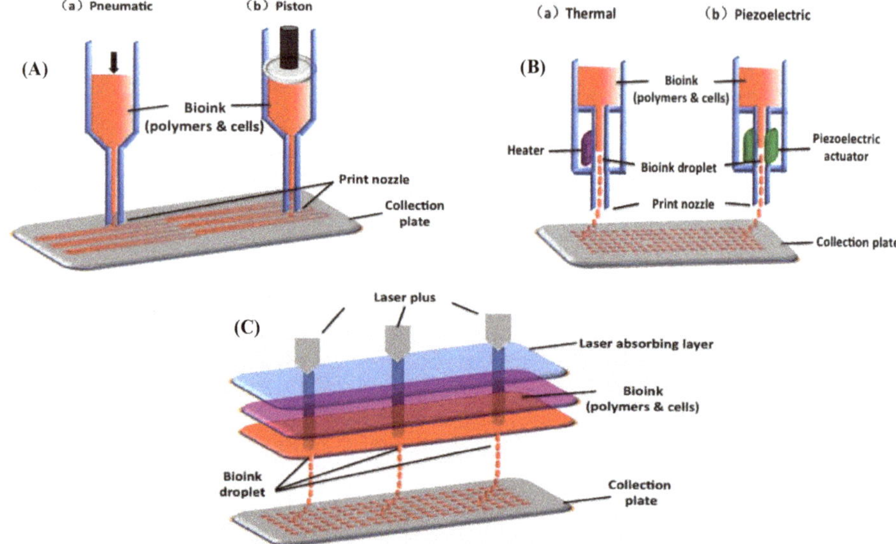

Figure 2. A schematic diagram of the three major types of organ 3D bioprinting technologies: (**A**) two-nozzle extrusion-based bioprinting (a: pneumatic; b: piston); (**B**) two-nozzle inkjet-based bioprinting (a: heater; b: piezoelectric actuator); (**C**) three-channel laser-assisted bioprinting. Image reproduced with permission from [26].

Thus, the purpose of organ 3D bioprinting is to design and manufacture bioartificial organs using polymeric materials, cells, bioactive agents, and advanced 3D printers. Defining and creating appropriate polymeric "bioinks" is a critical step in bioartificial organ 3D bioprinting [12–14]. Natural polymers, such as alginate, gelatin, hyaluronate, and chitosan, have been chosen as the preferable candidates for organ 3D bioprinting because of the specific properties, such as biocompatible, bioprintable, biodegradable, bioavailable, and biostable. Particularly, these natural polymers can be easily predesigned as the ECMs of each tissue in a natural organ.

3. Properties of Chitosan as a Natural Polymer

3.1. Resource of Chitosan

Chitin is one of the most abundant natural polymers in organism [27]. It is the second most abundant natural polymer next to cellulose, consisting of 2-acetamido-2-deoxy-d-glucose through a (1-4) linkage, extracted from the shells of marine crustaceans, insects, or fungi. Thus, the structural formula of chitin is poly-(1-4)-*N*-acetyl-glucosamine. Because of its facility to generate long-chain polymer structure, chitin is vitally important for some biological structure formation, such as cell walls in fungi and yeast, and exoskeleton of many invertebrates in shrimps and crabs. The extractive product is white, slightly pearly luster, and translucent sheet solid. Because of its insolubility in water and most of the organic solvents, chitin has limited applications in biomedical fields.

Chitosan, a derivate of chitin, is a linear polysaccharide, or carbohydrate polymer, derived from partial deacetylation of natural chitin. The deacetylation of chitin is conducted by chemical hydrolysis in alkaline conditions, using concentrated alkali water, or enzymatic hydrolysis with chitin deacetylase [28–30]. As its origin chitin, chitosan is highly available in nature.

3.2. Physico-Chemical Properties of Chitosan

Chitosan is a linear carbohydrate polymer derived from chitin with a structural similarity to glycosaminoglycan, a component of ECMs (Figure 3) [31–38]. Its chemical name is (1,4) -2-amino-2-deoxy-beta-*D*-glucan, a copolymer of randomly located (1-4)-2-amino-2-deoxy-d-glucan (d-glucosamine) and (1-4)-2-acetamido-2-deoxy-d-glucan (*N*-acetyl d-glucosamine) units. The number of amino groups as a ratio between d-glucosamine to the sum of d-glucosamine and *N*-acetyl d-glucosamine is indicated as a deacetylation degree (DD) and ordinarily should be larger than 60%. Some chitosans (e.g., 50% deacetylation degree) are water-soluble, while most of the chitosans are acid-soluble. Thus, chitosan can be recognized as a semi-natural positively charged polysaccharide at acidic conditions. Chitosan molecules can be modified through changing the functional groups, such as OH, and NH_2 by -$COCH_3$, -CH_3, -CH_2COOH, -SO_3H, -$PO(OH)_2$, etc.

CH₂OH
O
OH
O
NHCOCH₃
n
deacylate
CH₂OH
O
OH
O
NH₂
n
Chitin
Chitosan
CH₂OR₁
O
OR₂
O
NHR₃
n
Derivatives of chitin and chitosan

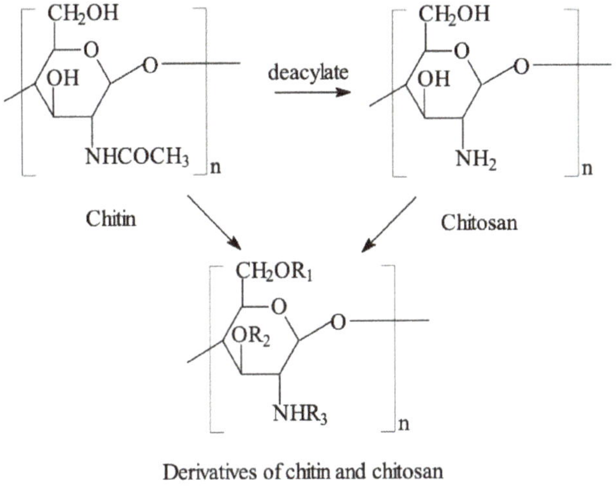

Figure 3. Chitosan generated by deacylation of chitin can be chemically modified through changing the functional groups, R_1, R_2, R_3 in the derivatives represents -$COCH_3$, -CH_3, -CH_2COOH, -SO_3H, -$PO(OH)_2$, etc. Image reproduced with permission from [31].

3.3. Prominent Characteristics of Chitosan for 3D Bioprinting

The biophysical characters of chitosan relay on a few factors, such as the molecular weight, DD, and purity of the molecules. Compared with animal derived natural polymers, such as collagen, gelatin, and fibrin, the degradation rate of chitosan is relatively slow. The biophysical properties of chitosan can be easily adjusted through changing the deacetylation rate of the original chitin and the molecular weight of the product.

Chitosan molecules possess several important properties, such as biocompatibility, biodegradability, antibacterial activity, non-antigenicity, and bioadsorbility. The solubility of chitosan in diluted acids is pH dependent through protonation of amino groups of the d-glucosamine residues. The availability of protonated amino groups enables chitosan to form complexes with metal ions, natural or synthetic anionic poly(acrylic acid) polymers, lipids, proteins, and deoxyribonucleic acid (DNA) [39–41]. The cationic feature of chitosan molecules favors the formation of gel particles through electrostatic interactions, with, e.g., sodium sulfate employed as a precipitant [39–41].

Stress should be given to the cationic feature of chitosan molecules. Few polymers in nature possess such special characters. The positive charged chitosan molecules can interact with hydrophobic components, giving rise to amphiphilic particles with great self-assembly and encapsulation capabilities. The polycationic nature of chitosan at a mild acidic condition allows the immobilization of negatively charged enzymes, proteins, and DNA for gene delivery and other composite biomaterial formation [42,43]. It has proven that appropriate interactions between chitosan molecules and drugs can produce expected pharmacological effect at the target site. The hydrophilic structure of chitosan promotes almost all cell types to adhere and proliferate on its scaffolds and makes chitosan a good candidate for tissue repair and organ 3D bioprinting.

4. Chitosan-Base Polymers in Tissue Repair and 3D Bioprinting

4.1. Antimicrobial Activities for Skin Regeneration

Chitosan and its derivations have shown a lot of distinctive characters in skin regeneration with antimicrobial activities. The mechanisms of antibiosis for Gram positive and Gram-negative bacteria are different. The differences are closely related to the composition of the bacterial walls [44,45]. It is supposed that chitosan presents antimicrobial activities through effectively interacting with the outer membrane or cytoderm of the bacteria, probably, electrostatic attraction and osmotic pressure taken upon the dominant roles in the interactions.

In Gram-positive bacteria, such as staphylococcus and streptococcus, the cytoderm consists of peptidoglycan with a thickness of 20–80 nm. *N*-acetylmuramic acids in the cytoderm can absorb negatively charged teichoic acids through covalent linkages. At the same time, lipopolyteichoic acids form covalent bonds with the cytoplasmic membrane. When teichoic acids form a layer of high-density charges in the cytoderm, the cytoderm is strengthened and the ion transportation through the outer surface layers can be restrained. Compared with Gram-positive bacteria, the peptidoglycan stratum in the cytoderm of Gram-negative bacteria is relatively thin, which lies over the cytoplasmic membrane. This complexity is further covered by an additional outer envelope membrane, with the fundamental elements of lipoprotein and lipopolysaccharide.

In acidic condition (especially, pH <6), the quaternary ammonium groups (R-NH3+) in chitosan molecules can competitively integrate with the divalent metal ions, such as Ca^{2+} and Mg^{2+}. To maintain the balance of voltage, polyanions often passively bind with the cytoderm leave. This binding can lead to the cytoderm to be hydrolyzed and cause the leakage of intracellular components [44,45]. The hydrolysis of peptidoglycans also results in an increased electrical interaction, leading to the enhancement of solute conductivity and cytoplasmic β-galactosidase release in the cell suspensions [46–49].

Thus, the reason of chitosan affecting the permeability of the outer cytoderm is through forming an ionic type of bonding and preventing the intracellular transport of nutrients into the cells. This type of tunnel also increases the internal osmotic pressure, leading to the apoptosis of cells because of the lack of nutrients [50]. Some other studies have shown that chitosan can penetrate the multilayered (murein cross-linked) bacterial walls as well as the cytoplasmic membrane. Chitosan destroys the bacterial cells by binding to the DNA which prevents DNA transcription and interrupts protein and m-ribonucleic acid (mRNA) synthesis [50]. The destructiveness is highly dependent on the capability of chitosan to penetrate the multilayered cell wall and the cytoplasmic membrane. Nevertheless, chitosan has generally been regarded as a membrane disruptor rather than a penetrator during the antimicrobial activities.

Skin regeneration is a complex procedure that contains four dynamic phases—hemostasis, inflammation, proliferation, and tissue remodeling [51]. This dynamic process involves vascular stimulators, ECM components, soluble factors, and various cells. The treatment of skin injuries needs to ensure a very high degree of protection, strong anti-inflammatory effect, and minimum scar formation.

Chitosan solutions can be made into porous scaffolds with intrinsic antimicrobial properties. The antimicrobial effect of the porous scaffolds can be further promoted through adding other antimicrobial agents for wound healing. In one study, a chitosan-cordycepin hydrogel was prepared via adsorbing negatively charged cordycepin onto the positively charged chitosan molecules without adding any cross-linking agent [52]. In another study, polyethylene terephthalate was 3D printed with a chitosan solution [53]. Each layer of the textile polyethylene terephthalate-chitosan was loaded with chlorhexidine. The stability of the 3D printed porous scaffold was enhanced by a heating system, which also extended the delivery time of chlorhexidine up to seven weeks. Chitosan with other natural polymers can be coprinted into asymmetric membranes, and normally, the lower layer can directly contact the damaged skin [54]. The 3D printed membranes presented efficient antimicrobial capability against methicillin resistant staphylococcus aureus (MRSA) strains during the skin repair processes using a mouse model. The skin repair effect is similar to the commercially available products [55].

In addition to antimicrobial function, chitosan participates in all phases of skin regeneration in many ways. Chitosan molecules can effectively promote the migration of neutrophils, increasing the secretion of IL-8, a potent neutrophil chemokine [56]. This reaction is in correlation with the level of *N*-acetylation [57]. Moreover, chitosan can affect the expression of growth factors by increasing transforming growth factor-1 (TGF-1) expression in early post-injury phase and decreasing it in later stages through binding themselves to anionic growth factors [58,59]. Especially, chitosan molecules with high DD stimulate the proliferation of dermal fibroblasts, allowing fibrous tissue formation and re-epithelialization [60,61]. These unique properties of chitosan molecules make them favorable candidates in skin regeneration.

4.2. Hemostatic Activity for Wound Healing

Some chitosan molecules with specific molecular weight and DD demonstrate powerful hemostatic capability, which is independent of the coagulation pathway of the host [62–64]. The amine groups in the chitosan molecules can interact directly with coagulation factors, promoting the initiation of coagulation. When the DD of chitosan is 68.36%, chitosan molecules in a solution tend to form mesh-like structures and act with vascular components directly. Whereas higher DD results in stronger hydrogen bonds and crystalline structures within chitosan chains that have limited interaction with red blood cells [65]. The interactions of chitosan molecules with polyelectrolytes can be enhanced when the molecular weight of chitosan is increased, so as to the procoagulation processes [66,67]. There are several chitosan-containing hemostatic products, such as Celox®, HemCon®, Axiostat®, Chitoflex®, and Chitoseal®, available in the market, which have been approved by the Food and Drug Administration of the United States (FDA) [68].

It is found that 3D bioprinted chitosan/collagen films are useful in wound healing. The host tissues had anaphylaxis reactions to allogenic source collagens. It is necessary to prepare more biocompatible chitosan/collagen substitutes for wound healing in the future. Human keratin-chitosan membrane produced through UV-crosslinking has shown the potential as wound dressing with improved mechanical properties [69]. Chitosan-chondroitin sulfate-based polyelectrolyte complex has shown strong hemostatic capability beside antimicrobial effect for wound healing applications [70]. 3D printed chitosan with positive charged bioactive agents, such as growth factors and cytokines, can promote the wound healing capability. In an attempt, nanoparticles of chitosan generated by ionotropic gelation with tripolyphosphate were loaded with granulocyte-macrophage colony-stimulating factor (GM-CSF). The complexity was freeze-dried afterward, leading to the production of nanocrystalline cellulose–hyaluronic acid combination [71,72]. Polycaprolactone nanofibers loaded with chitosan NPs containing GM-CSF accelerated wound closure phenomenon [73]. 3D printing of chitosan combined with peptides presents the ability of wound closure as well. Bioprinting of cell-laden chitosan hydrogels, containing Ser-Ile-Lys-Val-Ala-Val-chitosan macromers can effectively induce various types of collagen expression, prompting angiogenesis with markers of TGF-1 [74]. Meanwhile, the inflammatory factors, which are not conducive for wound healing, such as TNF-α, IL-1β, and IL-6 mRNA in a mouse skin wound model were significantly inhibited [75]. In some other studies, chitosans were used to enhance the affinity of growth factors. A 3D printed chitosan scaffold containing heparin-like polysaccharide (2-*N*, 6-*O*-sulfated) demonstrated an enhanced capability to attract vascular endothelial cells and induce the secretion of growth factors because of the high sulfonation degree [76].

4.3. Three-Dimensional Constructs for Bone Rehabilitation

Traditionally, chitosan membrane is one of the commonly used biomaterials in biomedical and clinical applications. It can be prepared through various technologies, such as electrospinning, thermal induced phase separation and self-assembly. During electrospinning, chitosan fibers are deposited irregularly to form non-woven fibrous membranes. The physical structures of the non-woven fibrous membranes are similar to those of natural ECMs. The shortages of the non-woven fibrous membranes to be used as tissue engineering scaffolds are the small pore sizes and weak mechanical strengths. On the one side, the pore sizes of the fibrous membranes are too small to let cells grow in [77]. For example, Sajesh et al. prepared a chitosan fibrous membrane through electrospinning [78]. The tensile strength of the fibrous membrane was 10 MPa, the average pore size was 5 um, and the porosity was over 80%. Shalumon found that in a high chitosan-contenting membrane, the tensile strength was only 1.5 MPa, which was lower than that of a commonly clinically used bone regeneration membrane [79]. The tensile strength of chitosan fibrous membrane needs to be improved. The pore size of chitosan membranes prepared through chitosan molecule self-assembly is also small, and the diameter of the pores is easily affected by the concentration of the chitosan molecules, solution pH, temperature, and other factors. Some scholars used sodium chloride as pore-forming agent to obtain large pores in the chitosan membranes via thermal induced phase separation.

Current research shows that calcium phosphate, carbon nanotubes, and hydroxyapatite can increase the mechanical properties of the chitosan scaffolds to some degree [80]. For example, Matinfar et al. mixed chitosan and carboxymethyl cellulose (CMC), and reinforced with whisker-like biphasic and triphasic calcium phosphate fibers as bone repair scaffolds [81]. The composite chitosan/CMC were obtained by freeze drying. The composite scaffolds exhibited desirable microstructures with high porosity (61–75%) and interconnected pores in range of 35–200 μm. Addition of CMC to chitosan solution led to a significant improvement in the mechanical properties (up to 150%) but did not affect the water uptake ability and biocompatibility. The composite chitosan/CMC scaffolds reinforced with 50 wt% triphasic fibers were superior in terms of mechanical and biological properties and showed compressive strength and modulus of 150 kPa and 3.08 MPa, respectively, which is up to 300% greater than pure chitosan scaffolds. Bi et al. prepared a chitosan-containing composite scaffold and seeded with osteoblasts. It was found that a large number of osteoblasts adhered on the scaffold and proliferated inside the go-through pores [82]. When the chitosan-containing composite scaffold was implanted into rats with skull-parietal bone loss, new bone formed at the edge of the bone loss site and the center of the scaffold in 2 weeks. After five weeks' implantation, new bone mass was significantly higher than that of the blank control.

With the introduction of 3D printing technologies in tissue engineering, the physical, biochemical, and physiological properties of the 3D printed chitosan scaffolds can be greatly improved. In a 3D printed chitosan scaffolds, osteoblasts grew along the computer controlled go-through channels and formed trabecula structures. Meanwhile blood vessels are easy to form along the go-through channels with the addition of endothelial cells [83]. Emphasis should be given to those growth factors, i.e., polypeptides, that can bind to specific cell membrane receptors to control cell destiny and regulate cell functions. Osteoinductive growth factors include vascular endothelial factor (VEGF), bone morphogenetic protein (BMP), platelet-derived growth factor (PDGF), and TGF, etc. Kjalarsdttir et al. cultured mouse fibroblasts with BMP-2 encapsulated in chitosan microsphere through a ion cross-linking method. The results showed that the encapsulation rate of chitosan microspheres was over 80% with a slow growth factor releasing rate. The sustainable releasing time attained 30 days. When the rhBMP-2 adsorbed chitosan microspheres were compounded on collagen sponge scaffolds and implanted into rabbits with radial segmental defects, the rhBMP-2-adsorbed chitosan microsphere scaffolds had more new bone mass than that of the control rhBMP-2/collagen scaffolds 12 weeks after the implantation, indicating that chitosan microspheres as carriers could effectively maintain the biological activity of rhBMP-2 [84]. The chitosan molecules can inhibit the secretion of osteoclasts and promote the proliferation of osteoblasts, thus promoting the bone tissue repair effect. When the

VEGF-containing chitosan microspheres were implanted into rat peritoneal adipose tissues, two weeks later, the number of endothelial cells and erythrocytes in the rats was significantly higher than that of the controls. These results suggest that the chitosan-containing 3D scaffolds together with growth factors can effectively promote large bone repair rate [85].

In some other researches, the positive charge amino groups in the chitosan molecules can combine with negative charged DNA molecules to form nanoparticles through polyelectrolyte actions. Foreign genes can be transfected into the cells of the body and play some roles in the cell behaviors. For example, Zeng et al. prepared nanoparticles via the reaction of mercaptan-organized chitosan and recombinant plasmid polyelectrolyte. When the recombinant plasmids containing *BMP-4* and *VEGFR1* genes were implanted into rabbits with radius defect, the experimental group had faster bone defect repair speed and more new bone mass compared with the controls [86]. Similarly, chitosan/polyacrylic acid nanofibers had been used as effective carriers of DNA plasmids. These researches have elaborated the active roles of chitosan molecules in bone tissue rehabilitation processes at molecular and cellular levels.

For large bone repair, a series of pioneering work have been done by the corresponding author of this article herself before 2000 through chemical modification of chitosans (Figures 4–6) [31–36]. For example, several large bone repair materials have been created by adding phosphorylated chitin (P-chitin), phosphorylated chitosan (P-chitosan), and disodium (1→4)-2-deoxy-2-sulfoamino-β-*D*-glucopyranuronan (S-chitosan) as the additives of biodegradable calcium phosphate cement (CPC) systems. The large bone repair materials are biocompatible, bioabsorbable, osteoconductive and/or osteoinductive. In vitro and in vivo experiments have shown that the bone repair rates and effects are directly related to the functional groups on the chitosan-based molecules and polymer concentrations in the CPCs. There are many different bone repair manners with these materials: some new trabeculae form directly after body fluid infiltration of the implants (Figure 5) [33]; some new trabeculae form following chondrocytes disappearing around the implants (Figure 6) [32]; some new trabeculae form after fibroblast-like cells being swallowed up [36]. The biodegradation rates of the materials have negative relationships with the P-chitin, P-chiosan, and S-chitosan contents. Most of the low concentration samples degrade in 16 weeks. While the high concentration samples disappear around 22 weeks. The fastest bone repair rates comes from those samples containing low concentrations of P-chitin and P-chitosan. Especially, the P-chitosan contained CPCs possess excellent biocompatibilities which can be transferred to trabeculae straightly after body fluid infiltration without any vise reactions or adverse effects, such as hematoma, inflammation, fibrous encapsulation, tissue necrosis, and excessive growth. The degradation rates of P-chitin and P-chitosan contained samples can be adjusted to match the ingrowth speeds of new trabeculae. A mild foreign-body reaction appears in the high P-chitin content samples during the early implantation stages which do not impair the final bone repair effects.

Later in 2003, the chitosan-based polymers have been 3D printed into hybrid large bone repair scaffolds with synthetic polymers, such as poly(lactic acid-co-glycolic acid) (PLGA), and mineral salts, such as calcium phosphate (TCP). Some of them have multiple functions, such as promoting osteoblast growth and inhibiting osteoclast activity (Figure 7) [87]. Different CAD models have been utilized to manufacture the hybrid scaffolds. Optimal fabrication parameters have been systematically studied through manipulating the processing materials. Furthermore, the microscopic structures, water absorbability, and mechanical properties of the hybrid scaffolds can be easily adjusted through adding different amount of P-chitin, P-chitosan, and S-chitosan. These hybrid scaffolds have been proven to be promising candidates for large hard tissue and organ manufacturing and restoration with later animal tests.

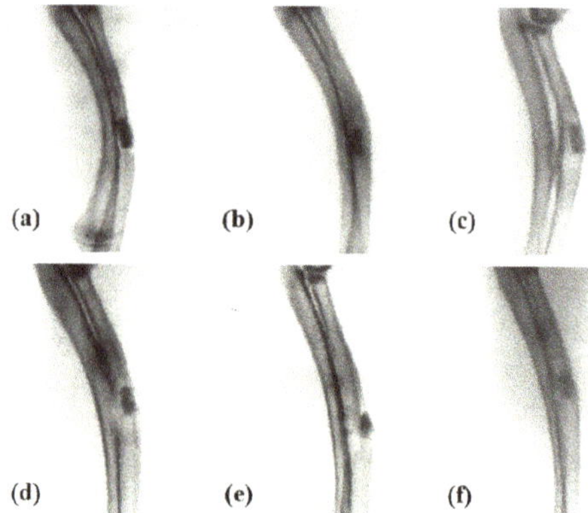

Figure 4. X-ray radiographs of large bone repair in rabbits with phosphorylated chitosan (P-chitosan) containing biodegradable calcium phosphate cements (CPCs): (**a**) 1 week (0.12 g/mL P-chitosan); (**b**) 4 weeks (0.14 g/mL P-chitosan); (**c**) 12 weeks (0.12 g/mL P-chitosan); (**d**) 12 weeks (0.07 g/mL P-chitosan); (**e**) 12 weeks (0.02 g/mL P-chitosan); (**f**) 22 weeks (0.12 g/mL P-chitosan). Image reproduced with permission from [33].

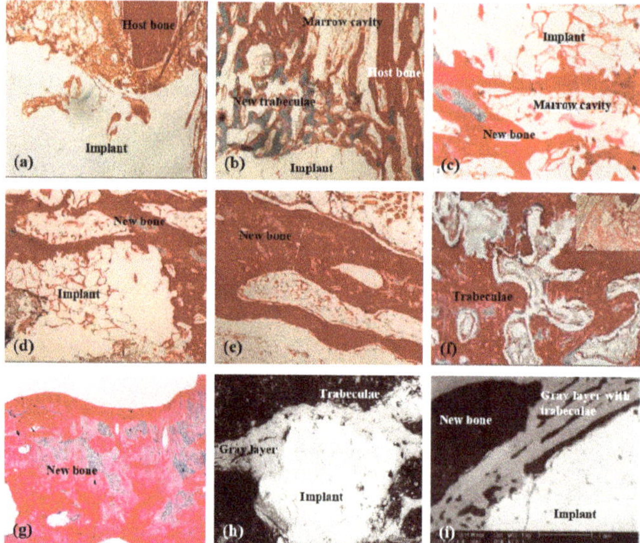

Figure 5. Tissue responses to different samples containing different concentrations of phosphorylated chitosan (P-chitosan) at different time points after implantation: (**a**) 1 week (0.12 g/mL P-chitosan), Masson Trichroism (M-T) staining; (**b**) 4 weeks (0.12 g/mL P-chitosan), M-T staining; (**c–e**) 12 weeks (0.02 g/mL P-chitosan); (**f**) 12 weeks (0.12 g/mL and 0.05 g/mL P-chitosan respectively), M-T and haematoxylin-eosin staining; (**g**) 22 weeks (0.12 g/mL P-chitosan) M-T staining; (**h**) 12 weeks (0.02 g/mL P-chitosan), a back scattered scanning electron microscopy (BSE) image; (**i**) 12 weeks (0.02 g/mL P-chitosan), a BSE image. Image reproduced with permission from [33].

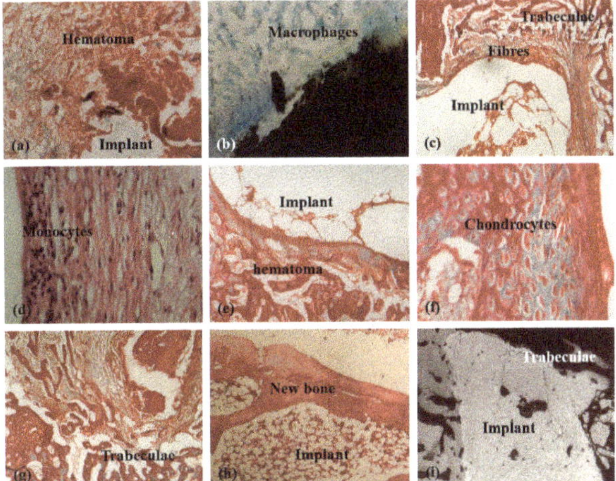

Figure 6. Tissue responses to different samples containing different concentrations of phosphorylated chitin (P-chitin) at different time points after implantation: (**a**,**b**) 1 week (0.14 g/mL P-chitin), Masson Trichroism (M-T) and Giemsa staining respectively; (**c**,**d**) 4 weeks (0.14 g/mL P-chitin), M-T and haematoxylin-eosin staining respectively; (**e**,**f**) 4 weeks (0.08 g/mL P-chitin); (**g**) 12 weeks (0.08 g/mL P-chitin) M-T staining; (**h**) 22 weeks (0.14 g/mL P-chitin); (**i**) 12 weeks (0.02 g/mL P-chitin), a back scattered scanning electron microscopy image. Image reproduced with permission from [32].

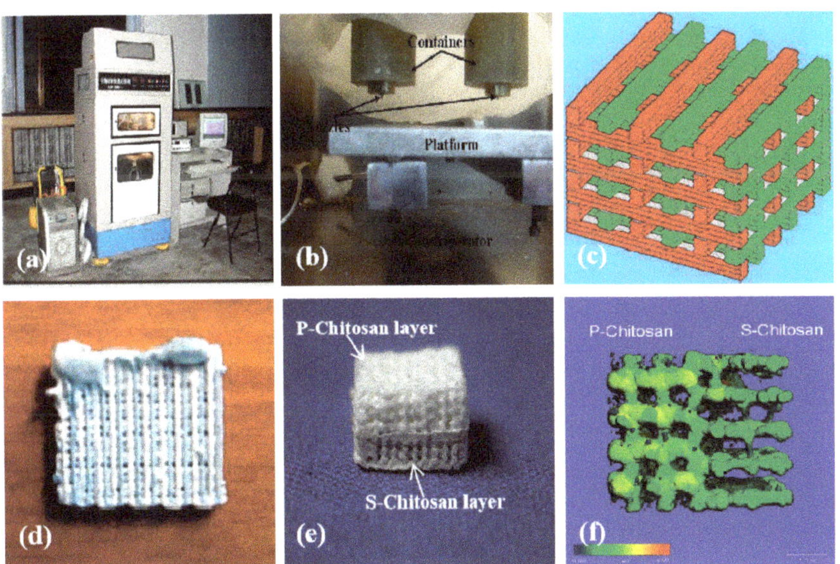

Figure 7. Graphical description of large bone repair scaffolds made in Tsinghua University, the corresponding author's laboratory in 2007: (**a**) a double-nozzle low-temperature 3D bioprinter; (**b**) working state of the double nozzles; (**c**) a grid computer-aided design (CAD) model containing two material systems; (**d**) a sample made from chitosan/gelatin and polyurethane; (**e**) a sample containing both P-chitosan and S-chitosan made via the double-nozzle low-temperature 3D bioprinter; (**f**) a computerized tomography of the 3D printed sample of (**e**). Image reproduced with permission from [87].

4.4. Cartilage Reconstruction

The treatments of cartilage degeneration and damaging are critical tasks in orthopedics. Articular cartilage injury may occur in sports, osteoarthritis, and many other situations. The common clinical therapeutic treatments include microfracture, mosaicplasty, autologous chondrocyte, and biomaterial implantation [88]. However, the vital limitation for cartilage regeneration is the insufficiency of vascular supply in the new cartilage tissues. To generate a graft with not only the capability to promote cartilage regeneration, but also to improve the avascular conditions, is the predominant goal of cartilage tissue reconstruction [89].

As stated above, chitosan is a natural polymer with a configurational similitude to sulfate glycosaminoglycans, and presents a similar microenvironment for the proliferation of chondrocyte, simulating ECM synthesis, and promoting chondrogenesis [90–96]. Compared with solitary alginate beads, the composite chitosan-alginate beads have shown enhanced chondrogenesis capacity when chondrocytes are embedded in. This can reduce the releasing of inflammatory cytokines, such as IL-6 and IL-8, and stimulate cartilage matrix component, such as hyaluronan and aggrecan, synthesis in vitro [97]. The derivates of chitosan have similar physiological activities and functions. For example, carboxymethyl-chitosan significantly reduced iNOS and IL-10 expression in a dose-dependent manner [98]. The addition of hyaluronic acid-chitosan to a pellet co-culture of the human infrapatellar fat pad (IPFP)-derived mesenchymal stem cells (MSCs) with osteoarthritic chondrocytes could increase chondrogenic differentiation [99].

From the molecular level, chitosan interacts with collagen through abundant electrostatic interactions between amino and sulfonate groups [100]. Compared with single component scaffolds, the freeze-dried type II collagen-chitosan composite scaffolds have better stiffness and ideal porous structures which are similar to natural cartilage ECMs [101]. It was reported that type II collagen-chitosan scaffolds combined with PLGA in bilayer forms can improve the mechanical and functional properties of the cartilage regeneration grafts [102]. Chitosan-silk fibroin blends can promote the cartilage regeneration efficiency [103,104]. One study found that the ratio of type II collagen to type I collagen of bovine chondrocytes cultured on 300-nm-diameter chitosan fibers was twice as high as that cultured on spongy scaffolds. It is noteworthy that the new macroporous 3D scaffolds prepared with freeze gel method (i.e., Cryogel) are attractive in biomedical fields [39–41]. The chitosan agarose gelatin Cryogel has super large pores (85–100 mm long) and good mechanical properties. The compression modulus of 5% Cryogel is about 44 kPa at 15% deformation [105]. When the chitosan-agarose-gelatin gel was used to repair subchondral cartilage defects in female New Zealand rabbits, there was no hypertrophic marker in the formation of hyaline cartilage around the fourth week after implantation. It is chitosan molecules that induce human bone marrow mesenchymal stem cells to differentiate into chondroid spheres by activating mTOR/S6K.

Until present, the widely used chitosan grafts for cartilage repair include cell-laden hydrogels, injectable solutions, and 3D printed scaffolds. Most of the cell-laden hydrogels contain both cellular (e.g., chondrocytes and undifferentiated progenitor chondrocytes) and bioactive molecular components (e.g., peptides, growth factors, and cytokines). These cell-laden hydrogels have low mechanical strength ($E \approx 200$ kPa) [88]. Intra-articular injection of the chitosan containing solutions can avoid the trauma of open surgery without affecting chondrocyte colonization and cartilage differentiation. However, the mechanical strengths of the injectable solutions are also very low to meet the clinical requirements.

Ideally, the biological properties of cartilage grafts should be capable of maintaining adult cell activities and inducing stem cell differentiation. From this point of view, the 3D printed constructs containing suitable biodegradable polymers and cell types are preferable [90–92]. In a previous study, a 3D-bioprinted chitosan scaffold attached with TGF-β and BMP-6 and seeded with human IPFP-MSCs demonstrated effective cell proliferation capacity with modified cartilage repair process [105].

It is expected that the 3D printed cartilage grafts hold the following prominent characteristics: (1) The selected biodegradable polymers are highly similar to natural cartilage ECMs, including the microstructures, physicochemical, and biochemical properties; (2) the incorporated cells are able to provide a temporary template to maintain the normal tissue function and synthesize new ECMs [92]; (3) the interconnecting porous structures are beneficial for cell migration and colonization [90]; (4) the 3D printed cell-laden constructs can be implanted in the articular cavity to cover the wound and promote cartilage repair [93].

4.5. Three-Dimensional-Bridge for Nerve Repair

Chitosan molecules have excellent neural biocompatibilities [106–108]. In vitro studies have shown that chitosan fibers or membranes can effectively promote nerve cell migration and proliferation. The survival time of hippocampal neurons and Schwann cells on chitosan films can be significantly prolonged. When a chitosan conduit was sutured outside a sciatic nerve gap of 10 mm or 15 mm, the motor and sensory functions of the never damaged rats could be totally recovered [109–112]. This chitosan conduit could reconstruct the long-distance peripheral nerve defect in diabetic rats and achieve similar recovery effects as autologous nerve transplantation [113–115]. Neurotrophic factors such as nerve growth factors could mix to the chitosan conduit through 3D bioprinting technology to promote peripheral nerve repair [116].

Similar results have been achieved through 3D printed nerve conduits made from both chitosan and synthetic polymers [117]. When chitosan was printed into the outer microporous tube and polyglycolic acid (PGA) as the inner guiding filler, the two-component artificial conduit could bridge and repair a 30 mm sciatic nerve defect in dogs with nerve continuity recovery and target muscle re-neurotization [117,118]. For example, a chitosan-PGA nerve graft could repair a 10-mm defect in rat sciatic nerve defect and maintain the continuity of the nerve tissue for 3 to 6 months. By the combination of neuronal supportive cells with the 3D bioprinted chitosan-containing constructs, the grafts could repair even larger nerve defects. A chitosan/polylactic acid-glycolic acid (PLGA)-based graft together with autologous bone marrow mesenchymal cells could repair 50 mm long sciatic nerve gaps in dogs [119]. Six months after implantation, the damaged nerves and motor functions were restored with innervated target muscles. The graft could effectively bridge a 60-mm sciatic nerve defect in dogs with the similar repair results to those of autotransplants [120].

Furthermore, transplantation of Schwann cells derived from dorsal root ganglion into a nerve repair graft constructed by PLGA/chitosan could increase the diameter and behavior area of axons and improve motor function after sciatic nerve injury in rats [121]. The degradation products of chitosan, i.e., chitosan oligosaccharide, or chitooligosaccharide, had protective effects against neurotoxicity. These degradation products could protect hippocampal neurons from apoptosis [122,123]. Additionally, chitooligosaccharides could increase the survival and proliferation of Schwann cells, enhance the formation of myelin in axons, and accelerate the release of neurotrophic factors, such as brain-derived neurotrophic factors and nerve growth factors [124]. Direct intravenous injection of chitooligosaccharide after common peroneal nerve injury in rabbits could activate the muscle tissues, and enhance the number of myelinated nerve fibers, as well as the thickness of myelin sheath. The cross-sectional area of tibial posterior muscle fibers was obviously augmented [125–127]. A chitooligosaccharide filled silica gel tube could repair a 10-mm sciatic nerve defect in rats and promote peripheral nerve regeneration. It is supposed that the beneficial effect of chitooligosaccharides is to establish an allowable microenvironment by stimulating the proliferation of Schwann cells, and increasing macrophage infiltration [128–131].

4.6. Hepatic Tissue and Organ Restoration

The liver is an important biochemical reactor in the human body. It is a complex inner organ with more than six cell types and abundant vascular and biliary networks, which makes the liver regeneration extremely difficult. Liver injury is harmful and sometimes fatal. At present, the worldwide seriously scanty of orthotopic liver donors has exacerbated the demand for new therapeutic treatment for acute and chronic liver failures [132]. Because the formation of thrombus can lead to obstruction and decrease the efficiency of blood transportation, the design of hierarchical vascular networks and anti-thrombotic extracellular components are essential aspects for liver 3D bioprinting.

In nature, hepatocytes exist in a complicated environment, surrounded with different types of ECMs. Hepatocytes are anchor-dependent cells, especially sensitive to surrounding environments for preserving their viability and proliferability [133–135]. Chitosan, as a kind of natural biomaterial with specific properties, has broad applications in liver tissue and organ construction and restoration. In some earlier studies, chitosan/collagen matrix generated in the *N*-hydroxysuccinimide (NHS) buffer system using cross-linking agent 1-ethyl-3-(3-dimethylaminopropyl) carbodiimide (EDC) had the potential for liver tissue regeneration [37]. The EDC cross-linked chitosan/collagen matrix presented moderate mechanical properties, excellent biocompatibility with hepatocytes. Chitosan-collagen-heparin matrix demonstrated a superior vascular biocompatibility for liver tissue construction. Chitosan modified with galactose residues could improve the adhesion of hepatocytes and maintain the viability of hepatocytes. Park et al. demonstrated that the specific interaction between asialoglycoprotein receptor (ASGPR) and galactose ligand of glycol chitosan (GC), led to the synthesis of new ECM for hepatocyte adhesion [136]. Chung et al. proposed the potential to improve the short-term viability of hepatocytes cultured on alginate/chitosan scaffolds [137]. Kim et al. reported a long-term enhancement of hepatocyte function in alginate/chitosan scaffolds [132]. Yu et al. 3D printed scaffold-free "tissue strands" as a "bioink" [138]. Co-culturing of hepatocytes with fibroblasts on an alginate/chitosan scaffold could boost the sphere formation speed. Li et al. reported that fructose coupled to porous chitosan scaffolds through the reaction of amino groups and aldehyde groups could improve the hepatic functions [139]. Fructose is a specific ligand of ASGPR in hepatocytes. The fructose-modified chitosan could induce hepatocyte aggregation, and specific metabolic activity of the artificial liver tissues to some degree. Especially, hepatocytes on ammonia-treated chitosan/collagen membranes demonstrated polar growth capabilities (Figure 8).

Over the past decade, chitosan has become one of the main components of "bioinks" for tissue and organ 3D bioprinting because of its similar biochemical properties to glycosaminoglycans (GAGs) in ECMs (Table 1) [140–160]. Liver 3D bioprinting is a manufacturing process with the ultimate goal to treat injured or failed livers. The main objective of liver 3D bioprinting is to develop biological liver substitutes or analogues in which patient plasma can circulate inside the vascularized hepatic tissues with metabolically active hepatocytes. An important aspect of 3D liver bioprinting is to select right cell types, such as primary hepatocytes, vascular stem cells, and hepatic stem cells. The patient-derived primary hepatocytes are a useful cell source for bioartificial liver 3D bioprinting. Many researchers have tried to optimize the cell survival environments to maintain the biochemical functions of the primary hepatocytes, so that they can carry out as many physiological functions as possible [133,134]. For example, compared with pure chitosan/collagen and chitosan/gelatin hydrogels, the chitosan/collagen/heparin and chitosan/gelatin/heparin hydrogels containing vascular cells, such as endothelial cells and smooth muscle cells, may have more potentials for liver 3D bioprinting with hierarchical vascular network construction [8,37,38]. The micro-structures of the chitosan/collagen/heparin and chitosan/gelatin/heparin hydrogels could provide anticoagulant functions before the inner surface of the vascular network is fully endothelialized. Basically, both the pure chitosan/collagen, chitosan/gelatin and chitosan/collagen/heparin, chitosan/gelatin/heparin hydrogels have large surface-volume ratio for hepatocytes to survive, since hepatocytes tend to adhere to specific substrate to migrate and proliferate. The micro-structures of the chitosan-containing molecules are beneficial to the transportation of nutrition and oxygen.

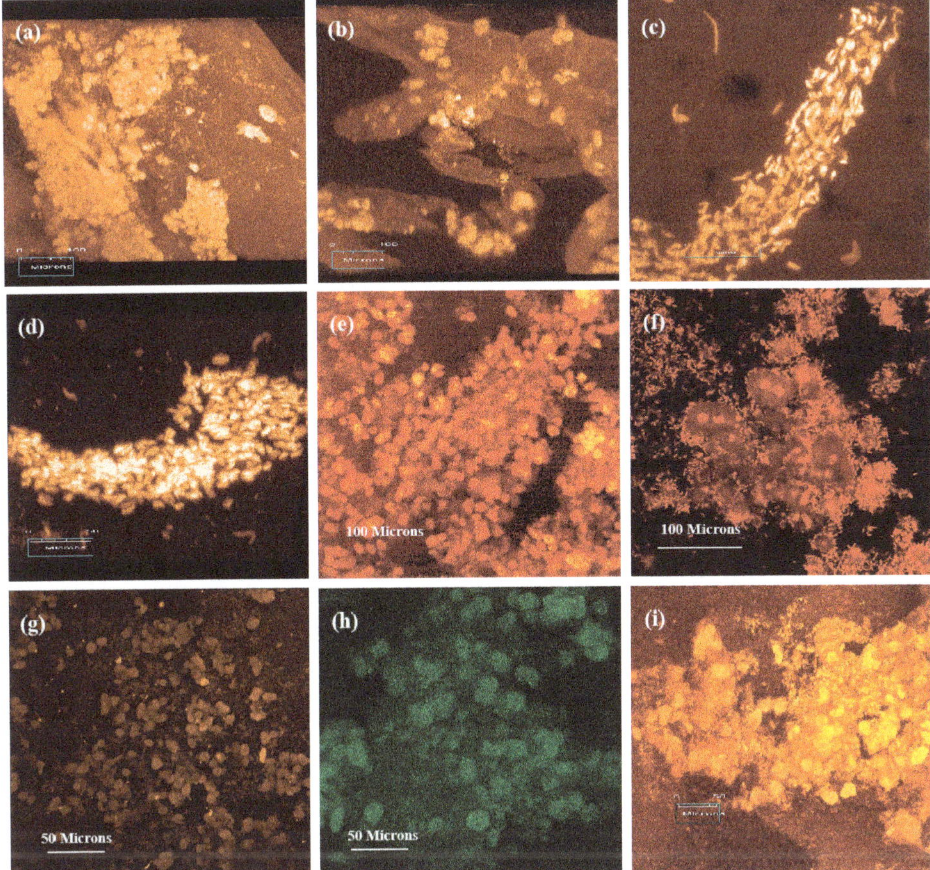

Figure 8. Laser scanning confocal microscope observations (propidium iodide staining) of hepatocytes seeded on the matrices of (**a**) ammonia treated collagen, (**b**) heparin sodium containing ammonia treated collagen, (**c**,**d**) ammonia treated chitosan/collagen, (**e**) heparin sodium containing ammonia treated chitosan/collagen, (**f**) sodium hyaluronate containing ammonia treated collagen/chitosan, (**g**,**h**,**i**) ammonia treated chitosan after 25 days of culture. Image reproduced with permission from [38].

Table 1. Chitosan containing "bioinks" for organ 3D bioprinting.

"Bioink" Formulation	3D Bioprinting Technique	Cross-Linking Method	Application	Ref
Chitosan/gelatin/hepatocytes	One nozzle extrusion-based 3D bioprinting	3% sodium tripolyphosphate (TPP)	Large scale-up hepatic tissue manufacturing	[154]
Chitosan/gelatin/alginate/hepatocytes and gelatin/alginate/fibrinogen/adipose -derived stem cells (ASCs)	Two nozzle extrusion-based 3D bioprinting	Triple crosslinking with TPP/$CaCl_2$/thrombin solutions after 3D bioprinting	Vascularized hepatic tissues with hierarchical branched vascular networks	[155]
Chitosan/sodium alginate (CS-SA) hydrogels	Rapid Prototyping (Fab@Home) printer	10% $CaCl_2$ (*w/v*) solution	Skin tissue regeneration	[156]
Oxidized hyaluronate (OHA)/glycol chitosan (GC)/adipic acid dihydrazide (ADH) hydrogels	3D bioprinter (Invivo®, Rokit, Korea)	Not require any post-gelation or additional cross-linking	Self-healing hydrogel system for cartilage regeneration	[157]
Chitosan-hydroxyapatite hydrogels (Chitosan-HA)	The Fab@Home™ (The Seraph Robotics, USA) open source RP platform Model 3	2% $CaCl_2$ (*v/v*) for 15 min	Bone regeneration	[158]
Chitosan/alginate hydrogel	One/two nozzle extrusion-based 3D bioprinting	$CaCl_2$ solution	Vessel-like tubular microfluidic channels	[159]
Polylactic acid-co-glycolic acid (PLGA)-gelatin/alginate/fibrinogen/ASCs -gelatin/chitosan/hepatocytes-gelatin/ hyaluronate/Schwann cells	Combined four-nozzle 3D bioprinting	Triple crosslinking with TPP/$CaCl_2$/thrombin solutions after 3D bioprinting	Vascularized and innervated liver tissue generating	[160]
Chitosan film enhanced chitosan nerve guides (CFeCNGs)	REAXON® Nerve Guide	Not required	Nerve regeneration	[115]

Chitosans, as additives of gelatin-based hydrogels have been frequently used as cell-laden "bioinks" for liver tissue and organ 3D bioprinting because of their good biocompatibility (i.e., no-toxicity, non-immunogenicity), chemical gelling capability (i.e., crosslinkability), moderate biodegradability, and ECM component simulation property. The first chitosan application in liver tissue 3D bioprinting is in 2003 (Figure 9) [154]. The obstacle for chitosan solutions to be printed alone is that their physical sol-gel transitions are too low to be below 0 °C, and it is difficult for the chitosan solutions to be printed with cells at room temperatures [143–145]. Physical blending of chitosan and gelatin solutions is necessary, endowing the composite gelatin/chitosan hydrogel with a higher phase transition temperature (i.e., sol-gel transition point ≈ 28 °C). The biophysical and biochemical characters of chitosan molecules are vital for generating chitosan-based polyelectrolytes for the layer-by-layer 3D deposition techniques.

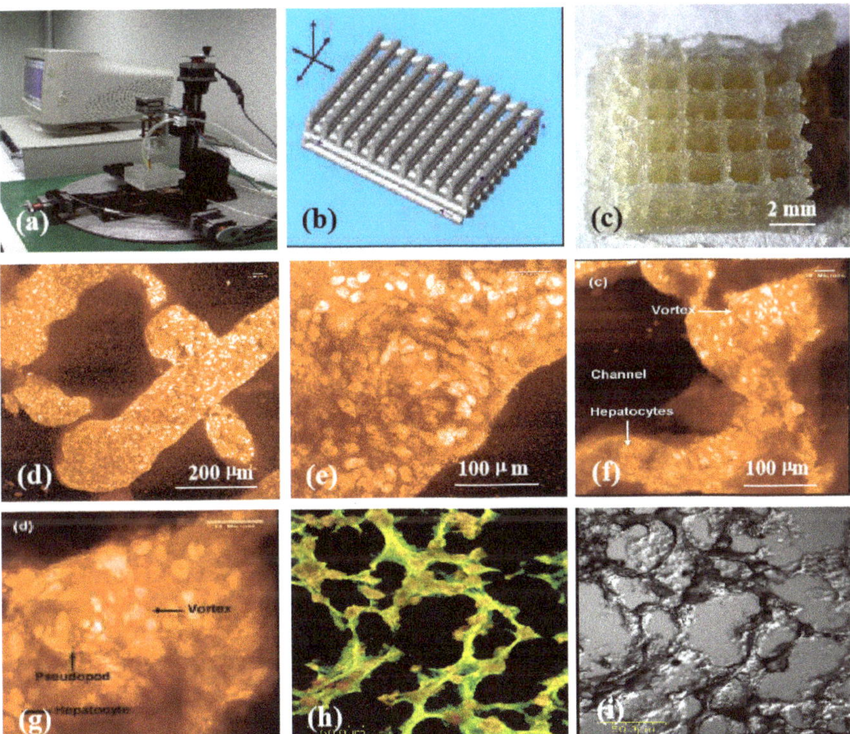

Figure 9. Three-dimensional (3D) printing of hepatocyte-laden chitosan/gelatin hydrogels and laser scanning confocal microscope observations of hepatocytes in the 3D printed chitosan/gelatin constructs with propidium iodide staining: (**a**) The 3D bioprinter made in the corresponding author's laboratory in 2003; (**b**) a grid computer-aided design (CAD) model; (**c**) a grid cell-laden 3D construct immediately after 3D bioprinting; (**d**,**e**) hepatocytes in the 3D printed chitosan/gelatin construct 1 month after in vitro culture; (**f**–**i**) hepatocytes in the 3D printed chitosan/gelatin construct 2 months after in vitro culture; (**e**,**g**) are the magnifications of (**d**) and (**f**) respectively; (i) a dark-field micrograph of (**h**). Image reproduced with permission from [154].

Since 2005, various chitosan-based composite "bioinks," such as chitosan/gelatin, chitosan/gelatin/alginate chitosan/gelatin/alginate/dextron-40 (glaycerol or dimethyl sulfoxide), have been explored in our laboratory through several extrusion-based 3D bioprinters (Figure 10) [155]. During the 3D bioprinting processes, the viscosity of the chitosan-based hydrogels depends largely on the composite polymer concentration, molecular weight, and cell density. These directly affect the 3D printing accuracy of the cell-laden constructs. After 3D bioprinting, the 3D printed constructs need to be stabilized using physical, chemical, and/or enzymic methods. The physical method includes thermosensitive transformation of the polymer molecules. While chemical and enzymic methods employ crosslinking agents or enzymes to crosslink or polymerize the incorporated polymer molecules. include chemical and enzymatic crosslinking.

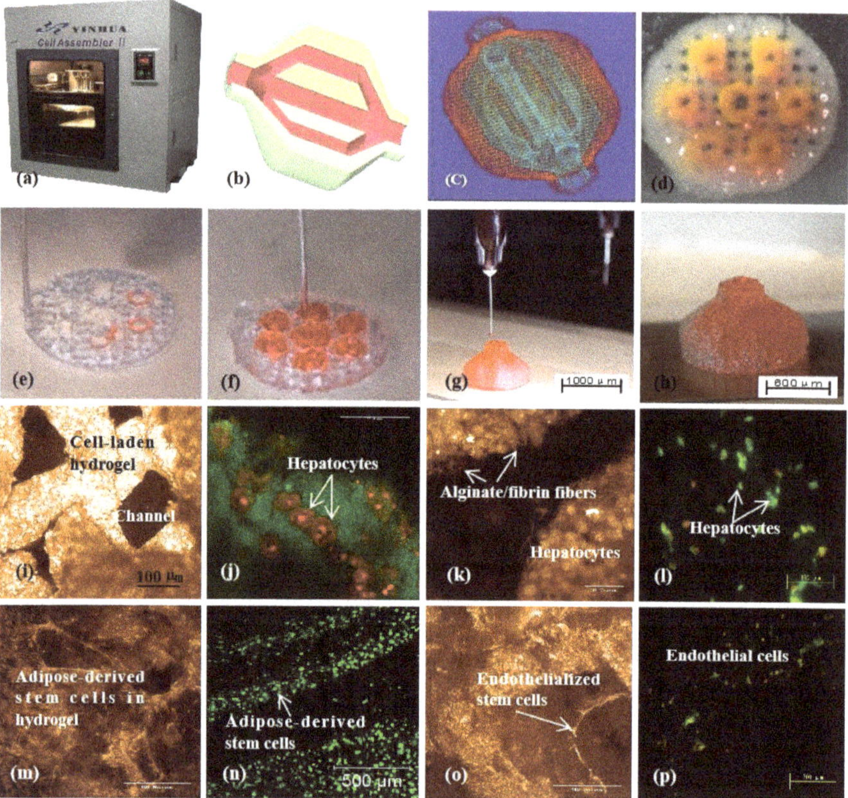

Figure 10. A large scale-up 3D printed vascularized liver tissue constructed through the double-nozzle 3D bioprinter created in Tsinghua University, the corresponding author's laboratory in 2006: (**a**) the double-nozzle extrusion-based 3D bioprinter; (**b**) a computer-aided design (CAD) model containing a hierarchical vascular network; (**c**) a CAD model containing the branched vascular network; (**d**) a few layers of the 3D bioprinted construct containing both adipose-derived stem cells (ASCs) encapsulated a gelatin/alginate/fibrin hydrogel and hepatocytes encapsulated in a gelatin/alginate/chitosan hydrogel; (**e**–**h**) 3D printing process of a semi-elliptical construct containing both ASCs and hepatocytes encapsulated in different hydrogels (i.e., gelatin/alginate/fibrin and gelatin/alginate/chitosan); (**i**–**l**) hepatocytes encapsulated in the gelatin/alginate/chitosan hydrogel after 3D bioprinting and different periods of in vitro cultures; (**m**–**p**) ASCs encapsulated in the gelatin/alginate/fibrin hydrogel after 3D bioprinting and different periods of in vitro cultures as well as growth factor inductions. Image reproduced with permission from [155].

Most of the stabilization processes contain two steps, both the thermosensitive physical and ionic chemical crosslinks. For example, the chitosan molecules can be chemically crosslinked using tripolyphosphate (TPP), oxidized dextran or other oxidized carbohydrates, 1,1,3,3-tetramethoxypropan, and genipin after a physical sol-gel transformation, ensuring the cell-laden constructs to be stable enough for being cultured in a liquid medium [42,43]. Further blending of chitosan with other natural polymers, such as fibrinogen, hyaluronan, endows the incorporated cells with more specific physiological functions [146–155].

Especially, the chemical crosslinking procedures for the 3D printed constructs can be changed depending on the composite polymer components. For example, a cell-laden chitosan/gelatin/alginate construct can be crosslinked by both TPP and $CaCl_2$. While, the cell-laden chitosan/gelatin/alginate/fibrinogen construct can be crosslinked by TPP, $CaCl_2$, and thrombin, respectively or in combination. During the chemical crosslinking procedures, the chitosan molecules are crosslinked by TPP. The alginate molecules are crosslinked by $CaCl_2$. Meanwhile the fibrinogen molecules are crosslinked by thrombin. The more crosslinks, the more stable of the 3D printed constructs. Ten years later, these classical "bioink" formulations and crosslinking strategies have been widely adapted by many other groups all over the world [146–155].

For complex liver 3D bioprinting, multiple networks, including vascular, neural, and biliary, should be enclosed (Figure 11) [160]. This is vital for this bionic liver to be connected to the host tissues with anti-suture and anti-stress properties. The chitosan containing ECM similar hydrogels with a large amount of water are vital for hepatocytes to anchor and survive [154,155]. Within these hydrogels, the 3D printed hepatocytes have the similar physiological functions, such as albumin, proteoglycan, and fibronectin, secretion. Meanwhile, the implantable bioartificial livers made from 3D bioprinting demonstrate the potential of permanent liver replacement and restoration.

5. Challenges and Perspectives

Like building "a nuclear plant," organ 3D bioprinting needs to face the selection and assembling of multiple polymeric materials, heterogeneous cell types, and other bioactive agents. Although significant progress has been made in vascularized and innervated organ 3D bioprinting over the last decade, some challenges remain in biomimicking each of the complex organs of human beings with special architectures, multi-cellular components, and physiological functions. The main challenges that remain in complex organ 3D bioprinting can be classified into the following three aspects: (1) Immunological rejections (or immune reactions) and other side effects from the 3D bioprinted biomaterials, especially living cells and polymeric hydrogels; (2) powerful 3D "bioprinters" that can recapitulate all the critical factors in a complex organ, such as the hierarchical biliary networks in the liver, the delicate lymphatic networks in the skin, and the multi-functional tubules in the kidney; (3) long-term growth and restoration capabilities of the bioartificial organs.

For patient with a failure organ, there are limited supply of autologous adult organ cells. Extensive donor site morbidity and complication may arise from heterogeneic cell transplantation and non-biocompatible polymer implantation. Cell viability in the 3D printed constructs directly affects the final organ growth and restoration results. The combination of chitosan-based polymers with autologous stem cells, growth factors, and other bioactive agents as predefined "bioinks" is an effective way to solve the risk of immune reactions, and teratoma formations in the bioartificial organs. Long-term stability of the 3D constructs needs to be further certified [161–166].

Beside the cell and polymer selection, the currently available 3D bioprinters are obviously cannot provide all the necessary capabilities for multiple cell type, polymeric material, and geometrical structure integration as those in a natural complex organ. It is challenging to develop new powerful "bioprinters" to print more cell types or stem cells/growth factors along with the selected polymeric materials for each solid organ construction with a full spectrum of physiological functions [167–170].

In the future, organ 3D bioprinting will play an essential role in many pertinent sciences and technologies, such as adult and stem cell biology, tissue science and engineering, drug screening and delivery, energy metabolism and detection. Especially, various sophisticated 3D bioprinters will be developed to recapitulate the macro- and micro-environments of natural organs. The prominent features of the sophisticated 3D bioprinters are the capabilities to integrate more structural, material, and physiological functions in an organic construct, and to enhance the cell loading and organizing efficiencies for multiple tissue formation, maturation, and coordination in a predefined construct [12–14,128–131,151–155].

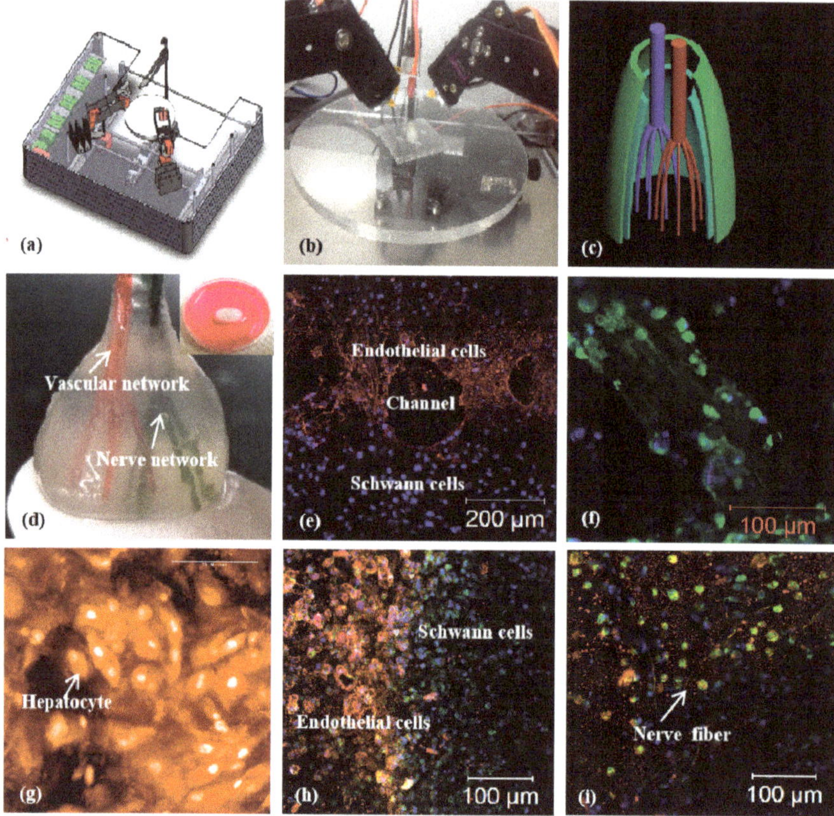

Figure 11. A combined four-nozzle organ three-dimensional (3D) bioprinting technology created in Tsinghua University, the corresponding author's laboratory in 2013 [160]: (**a**) equipment of the combined four-nozzle organ 3D bioprinter; (**b**) working state of the combined four-nozzle organ 3D printer; (**c**) a computer-aided design (CAD) model representing a large scale-up vascularized and innervated hepatic tissue; (**d**) a semi-ellipse 3D construct containing a poly (lactic acid-co-glycolic acid) (PLGA) overcoat, a hepatic tissue made from hepatocytes encapsulated in the gelatin/chitosan hydrogel, a branched vascular network with a fully confluent endothelialized adipose-derived stem cells (ASCs) on the inner surface of the cell-laden gelatin/alginate/fibrin hydrogel and a hierarchical never network made from Shwann cells encapsulated in the gelatin/hyaluronate hydrogel, the maximal diameter of the semi-ellipse construct can be adjusted from 1 mm to 2 cm according to the CAD model; (**e**) a cross section of (**d**), showing the interface of endothelialized ASCs and Schwann cells around a branched channel; (**f**) a large bundle of nerve fibers formed in the hierarchical never network of (**d**); (**g**) hepatocytes within the gelatin/chitosan hydrogel and underneath the PLGA overcoat; (**h**) a magnified picture of (**d**) showing the interface between the endothelialized ASCs and Schwann cells; (**i**) some thin nerve fibers near the hepatocyte-laden gelatin/chitosan hydrogel.

In the future, the 3D printed bioartificial organs will be used for a plenty of clinical purposes, such as failure organ restoration and cancer defect repair. One of the major advantages of the 3D printed bioartificial organs is that cells can be extracted from the patients themselves. Stem cells derived from patient themselves will be ideal cell sources. Immune rejection raised by the recipient's immune system can be excluded. By using the medical image, it is possible to reproduce the actual structure of the native organ of the patient.

In the future, the 3D printed bioartificial organs will offer a great potential for clinical applications, such as high throughput drug testing and screening, disease mechanism analysis and diagnosis, surgical intervention guidance and judgement [171–175]. Doctors and surgeons can practice and conduct experiments with these bioartificial organs to mimic the organ transplantation procedures. Scientists can test and use new drugs with these bioartificial organs to improve the therapeutic effects and eliminate the misdiagnosis complications.

In the future, there will be standardized global database for organ bank or library which allows access to all relevant medical images. Reverse organ manufacturing may become a hot issue for those intrinsic or accidental organ failures. Customize CAD models can either derive directly from the outcomes of a symmetrical organ or portrait. By using the advanced 3D bioprinters, the average life span of human beings will be significantly prolonged. The combination of architectural predesign, material optimization, medical imaging, advanced 3D bioprinting, and robotic surgery therefore will be a popular solution to most medical problems and should be a major research field in the forthcoming scientific and technological areas.

6. Conclusions

Chitosan, as a resourced natural molecule has been widely explored for biomedical applications. The prominent characteristics of chitosan such as non-toxic, biodegradable, antibiosis, and structural similarities to ECMs, make it a favorable candidate for a variety of tissue repair and 3D organ bioprinting applications. Especially, the structural similarities of chitosan with glycosaminoglycan are a favorable factor for tissue repair and organ 3D bioprinting. The 3D printed chitosan-containing porous scaffolds and cell-laden constructs have particular usages in promoting skin regeneration, wound healing, bone rehabilitation, cartilage reconstruction, nerve repair, and liver restoration. The biochemical and physiological properties of the chitosan-containing "bioinks" can easily be adjusted by changing the chitosan molecules in the hydrogels. It is expected that further study of chitosan molecules and their association with other polymers will reveal greater prospects of this unique polymer in complex organ manufacturing and clinical applications.

Author Contributions: S.L. designed and wrote the main content; X.T., J.F., H.T., and Q.A. contributed some detailed techniques, X.W. conceived, allocated, and revised the manuscript.

Funding: The work was supported by grants from the National Natural Science Foundation of China (NSFC) (Nos. 81571832 & 81571919 & 31600793 & 81271665), the 2017 Discipline Promotion Project of China Medical University (CMU) (No. 3110117049), the Key Research & Development Project of Liaoning Province (No. 2018225082), and the 2018 Scientist Partners of China Medical University (CMU) and Shenyang Branch of Chinese Academy of Sciences (CAS) (No. HZHB2018013).

Conflicts of Interest: The authors declare no conflict of interest. The founding sponsors had no role in the design of the study; in the collection, analyses, or interpretation of data; in the writing of the manuscript, and in the decision to publish the results.

References

1. Liu, F.; Chen, Q.; Liu, C.; Ao, Q.; Tian, X.; Fan, J.; Tong, H.; Wang, X. Natural polymers for organ 3D bioprinting. *Polymers* **2018**, *10*, 1278. [CrossRef] [PubMed]
2. Patrick, C.W. Adipose tissue engineering: The future of breast and soft tissue reconstruction following tumor resection. *Semin. Surg. Oncol.* **2000**, *19*, 302–311. [CrossRef]
3. Vasir, B.; Reitz, P.; Xu, G.; Sharma, A.; Bonner-Weir, S.; Weir, G.C. Effects of diabetes and hypoxia on gene markers of angiogenesis (HGF, cMET, uPA and uPAR, TGF-alpha, TGF-beta, bFGF and Vimentin) in cultured and transplanted rat islets. *Diabetologia* **2000**, *43*, 763–772. [CrossRef] [PubMed]

4. Dalal, A.R. Philosophy of organ donation: Review of ethical facets. *World J. Transpl.* **2015**, *5*, 44–51. [CrossRef] [PubMed]
5. Yelin, E.; Weinstein, S.; King, T. The burden of musculoskeletal diseases in the United States. *Semin. Arthritis Rheum.* **2016**, *46*, 259–260. [CrossRef] [PubMed]
6. Allemani, C.; Weir, H.K.; Carreira, H.; Harewood, R.; Spika, D.; Wang, X.S.; Bannon, F.; Ahn, J.V.; Johnson, C.J.; Bonaventure, A.; et al. Global surveillance of cancer survival 1995–2009: Analysis of individual data for 25,676,887 patients from 279 population-based registries in 67 countries (CONCORD-2). *Lancet* **2015**, *385*, 977–1010. [CrossRef]
7. Langer, R.; Vacanti, J.P. Tissue engineering. *Science* **1993**, *260*, 920–926. [CrossRef] [PubMed]
8. Wang, X.; Yan, Y.; Lin, F.; Xiong, Z.; Wu, R.; Zhang, R.; Lu, Q. Preparation and characterization of a collagen/chitosan/heparin matrix for an implantable bioartificial liver. *J. Biomater. Sci. Polym. Ed.* **2005**, *16*, 1063–1080. [CrossRef] [PubMed]
9. Moniaux, N.; Faivre, J. A reengineered liver for transplantation. *J. Hepatol.* **2011**, *54*, 386–387. [CrossRef] [PubMed]
10. Mao, A.S.; Mooney, D.J. Regenerative medicine: Current therapies and future directions. *Proc. Natl. Acad. Sci. USA* **2015**, *112*, 14452–14459. [CrossRef] [PubMed]
11. Hanauer, N.; Latreille, P.L.; Alsharif, S.; Banquy, X. 2D, 3D and 4D active compound delivery in tissue engineering and regenerative medicine. *Curr. Pharm. Des.* **2015**, *21*, 1506–1516. [CrossRef] [PubMed]
12. Wang, X.; Yan, Y.; Zhang, R. Rapid prototyping as a tool for manufacturing bioartificial livers. *Trends Biotechnol.* **2007**, *25*, 505–513. [CrossRef] [PubMed]
13. Wang, X. Intelligent freeform manufacturing of complex organs. *Artif. Organs* **2012**, *36*, 951–961. [CrossRef] [PubMed]
14. Wang, X.; Yan, Y.; Zhang, R. Recent trends and challenges in complex organ manufacturing. *Tissue Eng. Part B* **2010**, *16*, 189–197. [CrossRef] [PubMed]
15. Painter, P.C.; Coleman, M.M. *Fundamentals of Polymer Science: An Introductory Text*; Technomic Pub. Co.: Lancaster, PA, USA, 1997; p. 1. ISBN 1-56676-559-5.
16. Sperling, L.H. *Introduction to Physical Polymer Science*; Wiley: Hoboken, NJ, USA, 2006; p. 10. ISBN 0-471-70606-X.
17. Christensen, K.; Xu, C.; Chai, W.; Zhang, Z.; Fu, J.; Huang, Y. Freeform inkjet printing of cellular structures with bifurcations. *Biotechnol. Bioeng.* **2015**, *112*, 1047–1055. [CrossRef] [PubMed]
18. Norotte, C.; Marga, F.S.; Niklason, L.E.; Forgacs, G. Scaffold-free vascular tissue engineering using bioprinting. *Biomaterials* **2009**, *30*, 5910–5917. [CrossRef] [PubMed]
19. Benam, K.H.; Dauth, S.; Hassell, B.; Herland, A.; Jain, A.; Jang, K.-J.; Karalis, K.; Kim, H.J.; MacQueen, L. Mahmoodian, R. Engineered in vitro disease models. *Annu. Rev. Pathol. Mech. Dis.* **2015**, *10*, 195–262. [CrossRef] [PubMed]
20. Grigoryan, B.; Paulsen, S.J.; Corbett, D.C.; Sazer, D.W.; Fortin, C.L.; Zaita, A.J.; Greenfield, P.T.; Calafat, N.J.; Gounley, J.P.; Ta, A.H.; et al. Multivascular networks and functional intravascular topologies within biocompatible hydrogels. *Science* **2019**, *364*, 458–464. [CrossRef] [PubMed]
21. Labbaf, S.; Ghanbar, H.; Stride, E.; Edirisinghe, M. Preparation of multilayered polymeric structures using a novel four-needle coaxial electrohydrodynamic device. *Macromol. Rapid Commun.* **2014**, *35*, 618–623. [CrossRef] [PubMed]
22. Noor, N.; Shapira, A.; Edri, R.; Gal, I.; Wertheim, L.; Dvir, T. 3D printing of personalized thick and perfusable cardiac patches and hearts. *Adv. Sci.* **2019**, *6*, 1900344. [CrossRef] [PubMed]
23. Singh, S.; Afara, I.O.; Tehrani, A.H.; Oloyede, A. Effect of decellularization on the load-bearing characteristics of articular cartilage matrix. *Tissue Eng. Regen. Med.* **2015**, *12*, 294–305. [CrossRef]
24. Ye, L.; Zimmermann, W.-H.; Garry, D.J.; Zhang, J. Patching the heart cardiac repair from within and outside. *Circ. Res.* **2013**, *113*, 922–932. [CrossRef] [PubMed]
25. Lei, M.; Wang, X. Biodegradable polymers and stem cells for bioprinting. *Molecules* **2016**, *21*, 539. [CrossRef] [PubMed]
26. Wang, X.; Liu, C. 3D bioprinting of adipose-derived stem cells for organ manufacturing. In *Cutting-Edge Enabling Technology for Regenerative Medicine*; Khang, G., Ed.; Springer: Singapore, 2018; pp. 3–14.
27. Tharanathan, R.N.; Kittur, F.S. Chitin—The undisputed biomolecule of great potential. *Crit. Rev. Food Sci. Nutr.* **2003**, *43*, 61–87. [CrossRef] [PubMed]

28. Ghormade, V.; Pathan, E.K.; Deshpande, M.V. Can fungi compete with marine sources for chitosan production? *Int. J. Biol. Macromol.* **2017**, *104*, 1415–1421. [CrossRef] [PubMed]
29. Saranya, N.; Moorthi, A.; Saravanan, S.; Devi, M.P.; Selvamurugan, N. Chitosan and its derivatives for gene delivery. *Int. J. Biol. Macromol.* **2011**, *48*, 234–238. [CrossRef] [PubMed]
30. Venkatesan, J.; Anil, S.; Kim, S.K.; Shim, M.S. Chitosan as a vehicle for growth factor delivery: Various preparations and their applications in bone tissue regeneration. *Int. J. Biol. Macromol.* **2017**, *104*, 1383–1397. [CrossRef] [PubMed]
31. Wang, X.; Ma, J.; Wang, Y.; He, B. Structural characterization of phosphorylated chitosan and their applications as effective additives of calcium phosphate cements. *Biomaterials* **2001**, *22*, 2247–2255. [CrossRef]
32. Wang, X.; Ma, J.; Feng, Q.; Cui, F. Skeletal repair in of rabbits with calcium phosphate cements incorporated phosphorylated chitin reinforced. *Biomaterials* **2002**, *23*, 4591–4600. [CrossRef]
33. Wang, X.; Ma, J.; Wang, Y.; He, B. Bone repair in radii and tibias of rabbits with phosphorylated chitosan reinforced calcium phosphate cements. *Biomaterials* **2002**, *23*, 4167–4176. [CrossRef]
34. Wang, X.; Li, D.; Wang, W.; Feng, Q.; Cui, F.; Xu, Y.; Song, X. Covalent immobilization of chitosan and heparin on PLGA surface. *Int. J. Macromol.* **2003**, *33*, 95–100. [CrossRef]
35. Wang, X.; Ma, J.; Feng, Q.; Cui, F. The effects of S-chitosan on the physical properties of calcium phosphate cements. *J. Bioact. Compat. Polym.* **2003**, *18*, 45–57. [CrossRef]
36. Wang, X.; Ma, J.; Feng, Q.; Cui, F. In vivo testing of S-chitosan enhanced calcium phosphate cements. *J. Bioact. Compat. Polym.* **2003**, *18*, 259–271. [CrossRef]
37. Wang, X.; Li, D.; Wang, W.; Feng, Q.; Cui, F.; Xu, Y.; Song, X.; van der Mark, W. Crosslinked collagen/chitosan matrices for artificial livers. *Biomaterials* **2003**, *24*, 3213–3220. [CrossRef]
38. Wang, X.; Yan, Y.; Lin, F.; Xiong, Z.; Lin, F.; Wu, R.; Zhang, R.; Lu, Q. Preparation and evaluation of ammonia treated collagen/chitosan matrices for liver tissue engineering. *J. Biomed. Mater. Res. Part B Appl. Biomater.* **2005**, *70*, 91–98. [CrossRef] [PubMed]
39. Berillo, D.; Mattiasson, B.; Kirsebom, H. Cryogelation of chitosan using noble-metal ions: In situ formation of nanoparticles. *Biomacromolecules* **2014**, *15*, 2246–2255. [CrossRef] [PubMed]
40. Berillo, D.; Cundy, A. 3D-macroporous chitosan-based scaffolds with in situ formed Pd and Pt nanoparticles for nitrophenol reduction. *Carbohydr. Polym.* **2018**, *192*, 166–175. [CrossRef] [PubMed]
41. Berillo, D.; Elowsson, L.; Kirsebom, H. Oxidized dextran as crosslinker for chitosan cryogel scaffolds and formation of polyelectrolyte complexes between chitosan and gelatin. *Macromol. Biosci.* **2012**, *12*, 1090–1099. [CrossRef] [PubMed]
42. Akilbekova, D.; Shaimerdenova, M.; Adilov, S.; Berillo, D. Biocompatible scaffolds based on natural polymers for regenerative medicine. *Int. J. Biol. Macromol.* **2018**, *114*, 324–333. [CrossRef] [PubMed]
43. Lu, H.T.; Lu, T.W.; Chen, C.H.; Lu, K.Y.; Mi, F.L. Development of nanocomposite scaffolds based on biomineralization of N,O-carboxymethyl chitosan/fucoidan conjugates for bone tissue engineering. *Int. J. Biol. Macromol.* **2018**, *120*, 2335–2345. [CrossRef] [PubMed]
44. Young, D.H.; Kauss, H. Release of Calcium from Suspension-Cultured Glycine max Cells by Chitosan, Other Polycations, and Polyamines in Relation to Effects on Membrane Permeability. *Plant. Physiol.* **1983**, *73*, 698–702. [CrossRef] [PubMed]
45. Chung, Y.C.; Chen, C.Y. Antibacterial characteristics and activity of acid-soluble chitosan. *Bioresour. Technol.* **2008**, *99*, 2806–2814. [CrossRef] [PubMed]
46. Kong, M.; Chen, X.G.; Liu, C.S.; Liu, C.G.; Meng, X.H.; Yu le, J. Antibacterial mechanism of chitosan microspheres in a solid dispersing system against E. *coli. Colloids Surf. B Biointerfaces* **2008**, *65*, 197–202. [CrossRef] [PubMed]
47. Je, J.Y.; Kim, S.K. Chitosan derivatives killed bacteria by disrupting the outer and inner membrane. *J. Agric. Food Chem.* **2006**, *54*, 6629–6633. [CrossRef] [PubMed]
48. Liu, H.; Du, Y.; Wang, X.; Sun, L. Chitosan kills bacteria through cell membrane damage. *Int. J. Food Microbiol.* **2004**, *95*, 147–155. [CrossRef] [PubMed]
49. Vishu Kumar, A.B.; Varadaraj, M.C.; Gowda, L.R.; Tharanathan, R.N. Characterization of chito-oligosaccharides prepared by chitosanolysis with the aid of papain and Pronase, and their bactericidal action against Bacillus cereus and Escherichia coli. *Biochem. J.* **2005**, *391*, 167–175. [CrossRef] [PubMed]
50. Rabea, E.I.; Badawy, M.E.; Stevens, C.V.; Smagghe, G.; Steurbaut, W. Chitosan as antimicrobial agent: Applications and mode of action. *Biomacromolecules* **2003**, *4*, 1457–1465. [CrossRef] [PubMed]

51. Rousselle, P.; Montmasson, M.; Garnier, C. Extracellular matrix contribution to skin wound re-epithelialization. *Matrix Biol.* **2019**, 12–26. [CrossRef] [PubMed]
52. Ueno, H.; Yamada, H.; Tanaka, I.; Kaba, N.; Matsuura, M.; Okumura, M.; Kadosawa, T.; Fujinaga, T. Accelerating effects of chitosan for healing at early phase of experimental open wound in dogs. *Biomaterials* **1999**, *20*, 1407–1414. [CrossRef]
53. Park, C.J.; Gabrielson, N.P.; Pack, D.W.; Jamison, R.D.; Wagoner Johnson, A.J. The effect of chitosan on the migration of neutrophil-like HL60 cells, mediated by IL-8. *Biomaterials* **2009**, *30*, 436–444. [CrossRef] [PubMed]
54. Baxter, R.M.; Dai, T.; Kimball, J.; Wang, E.; Hamblin, M.R.; Wiesmann, W.P.; McCarthy, S.J.; Baker, S.M. Chitosan dressing promotes healing in third degree burns in mice: Gene expression analysis shows biphasic effects for rapid tissue regeneration and decreased fibrotic signaling. *J. Biomed. Mater. Res. A* **2013**, *101*, 340–348. [CrossRef] [PubMed]
55. Tsai, C.W.; Chiang, I.N.; Wang, J.H.; Young, T.H. Chitosan delaying human fibroblast senescence through downregulation of TGF-beta signaling pathway. *Artif. Cells Nanomed. Biotechnol.* **2018**, *46*, 1852–1863. [PubMed]
56. Howling, G.I.; Dettmar, P.W.; Goddard, P.A.; Hampson, F.C.; Dornish, M.; Wood, E.J. The effect of chitin and chitosan on the proliferation of human skin fibroblasts and keratinocytes in vitro. *Biomaterials* **2001**, *22*, 2959–2966. [CrossRef]
57. Hamilton, V.; Yuan, Y.; Rigney, D.A.; Puckett, A.D.; Ong, J.L.; Yang, Y.; Elder, S.H.; Bumgardner, J.D. Characterization of chitosan films and effects on fibroblast cell attachment and proliferation. *J. Mater. Sci. Mater. Med.* **2006**, *17*, 1373–1381. [CrossRef] [PubMed]
58. Song, R.; Zheng, J.; Liu, Y.; Tan, Y.; Yang, Z.; Song, X.; Yang, S.; Fan, R.; Zhang, Y.; Wang, Y. A natural cordycepin/chitosan complex hydrogel with outstanding self-healable and wound healing properties. *Int. J. Biol. Macromol.* **2019**, *134*, 91–99. [CrossRef] [PubMed]
59. Aubert-Viard, F.; Mogrovejo-Valdivia, A.; Tabary, N.; Maton, M.; Chai, F.; Neut, C.; Martel, B.; Blanchemain, N. Evaluation of antibacterial textile covered by layer-by-layer coating and loaded with chlorhexidine for wound dressing application. *Mater. Sci. Eng. C Mater. Biol. Appl.* **2019**, *100*, 554–563. [CrossRef] [PubMed]
60. Alves, P.; Santos, M.; Mendes, S.; Miguel, S.P.; de Sá, K.D.; Cabral, C.S.D.; Correia, I.J.; Ferreira, P. Photocrosslinkable nanofibrous asymmetric membrane designed for wound dressing. *Polymers* **2019**, *11*, 653. [CrossRef] [PubMed]
61. Shah, A.; Buabeid, A.M.; Arafa, E.A.; Hussain, I.; Li, L.; Murtaza, G. The wound healing and antibacterial potential of triple-component nanocomposite (chitosan-silver-sericin) films loaded with moxifloxacin. *Int. J. Pharm.* **2019**, *564*, 22–38. [CrossRef] [PubMed]
62. Khan, M.A.; Mujahid, M. A review on recent advances in chitosan based composite for hemostatic dressings. *Int. J. Biol. Macromol.* **2019**, *124*, 138–147. [CrossRef] [PubMed]
63. Hu, Z.; Lu, S.; Cheng, Y.; Kong, S.; Li, S.; Li, C.; Yang, L. Investigation of the effects of molecular parameters on the hemostatic properties of chitosan. *Molecules* **2018**, *23*, 3147. [CrossRef] [PubMed]
64. Yang, J.; Tian, F.; Wang, Z.; Wang, Q.; Zeng, Y.J.; Chen, S.Q. Effect of chitosan molecular weight and deacetylation degree on hemostasis. *J. Biomed. Mater. Res. B Appl. Biomater.* **2008**, *84*, 131–137. [CrossRef] [PubMed]
65. Guo, X.; Sun, T.; Zhong, R.; Ma, L.; You, C.; Tian, M.; Li, H.; Wang, C. Effects of Chitosan oligosaccharides on human blood components. *Front. Pharmacol.* **2018**, *9*, 1412. [CrossRef] [PubMed]
66. Klokkevold, P.R.; Fukayama, H.; Sung, E.C.; Bertolami, C.N. The effect of chitosan (poly-N-acetyl glucosamine) on lingual hemostasis in heparinized rabbits. *J. Oral Maxillofac. Surg.* **1999**, *57*, 49–52. [CrossRef]
67. Hattori, H.; Ishihara, M. Changes in blood aggregation with differences in molecular weight and degree of deacetylation of chitosan. *Biomed. Mater.* **2015**, *10*, 015014. [CrossRef] [PubMed]
68. Hu, Z.; Zhang, D.Y.; Lu, S.T.; Li, P.W.; Li, S.D. Chitosan-based composite materials for prospective hemostatic applications. *Mar. Drugs* **2018**, *16*, 273. [CrossRef] [PubMed]
69. Lin, C.W.; Chen, Y.K.; Lu, M.; Lou, K.L.; Yu, J. Photo-crosslinked keratin/chitosan membranes as ptential wound dressing materials. *Polymers* **2018**, *10*, 987. [CrossRef] [PubMed]
70. Sharma, S.; Swetha, K.L.; Roy, A. Chitosan-chondroitin sulfate based polyelectrolyte complex for effective management of chronic wounds. *Int. J. Biol. Macromol.* **2019**, *132*, 97–108. [CrossRef] [PubMed]

71. Koukaras, E.N.; Papadimitriou, S.A.; Bikiaris, D.N.; Froudakis, G.E. Insight on the formation of chitosan nanoparticles through ionotropic gelation with tripolyphosphate. *Mol. Pharm.* **2012**, *9*, 2856–2862. [CrossRef] [PubMed]
72. Karimi Dehkordi, N.; Minaiyan, M.; Talebi, A.; Akbari, V.; Taheri, A. Nanocrystalline cellulose-hyaluronic acid composite enriched with GM-CSF loaded chitosan nanoparticles for enhanced wound healing. *Biomed. Mater.* **2019**, *14*, 035003. [CrossRef] [PubMed]
73. Tanha, S.; Rafiee-Tehrani, M.; Abdollahi, M.; Vakilian, S.; Esmaili, Z.; Naraghi, Z.S.; Seyedjafari, E.; Javar, H.A. G-CSF loaded nanofiber/nanoparticle composite coated with collagen promotes wound healing in vivo. *J. Biomed. Mater. Res. A* **2017**, *105*, 2830–2842. [CrossRef] [PubMed]
74. Chen, X.; Zhang, M.; Chen, S.; Wang, X.; Tian, Z.; Chen, Y.; Xu, P.; Zhang, L.; Zhang, L.; Zhang, L. Peptide-modified chitosan hydrogels accelerate skin wound healing by promoting fibroblast proliferation, migration, and secretion. *Cell Transplant.* **2017**, *26*, 1331–1340. [CrossRef] [PubMed]
75. Chen, X.; Fu, W.; Cao, X.; Jiang, H.; Che, X.; Xu, X.; Ma, B.; Zhang, J. Peptide SIKVAV-modified chitosan hydrogels promote skin wound healing by accelerating angiogenesis and regulating cytokine secretion. *Am. J. Transl. Res.* **2018**, *10*, 4258–4268. [PubMed]
76. Yu, Y.; Chen, R.; Sun, Y.; Pan, Y.; Tang, W.; Zhang, S.; Cao, L.; Yuan, Y.; Wang, J.; Liu, C. Manipulation of VEGF-induced angiogenesis by 2-N, 6-O-sulfated chitosan. *Acta Biomater.* **2018**, *71*, 510–521. [CrossRef] [PubMed]
77. Zhang, X.Y.; Chen, Y.P.; Han, J.; Mo, J.; Dong, P.F.; Zhuo, Y.H.; Feng, Y. Biocompatiable silk fibroin/carboxymethyl chitosan/strontium substituted hydroxyapatite/cellulose nanocrystal composite scaffolds for bone tissue engineering. *Int. J. Biol. Macromol.* **2019**, *136*, 1247–1257. [CrossRef] [PubMed]
78. Sajesh, K.M.; Jayakumar, R.; Nair, S.V.; Chennazhi, K.P. Corrigendum to "biocompatible conducting chitosan/polypyrrole-alginate composite scaffold for bone tissue engineering". *Int. J. Biol. Macromol.* **2013**, *62*, 465–471. [CrossRef] [PubMed]
79. Shalumon, K.T.; Anulekha, K.H.; Chennazhi, K.P.; Tamura, H.; Nair, S.V.; Jayakumar, R. Corrigendum to "Fabrication of chitosan/poly(caprolactone) nanofibrous scaffold for bone and skin tissue engineering". *Int. J. Biol. Macromol.* **2011**, *48*, 571–576. [CrossRef] [PubMed]
80. Matinfar, M.; Mesgar, A.S.; Mohammadi, Z. Evaluation of physicochemical, mechanical and biological properties of chitosan/carboxymethyl cellulose reinforced with multiphasic calcium phosphate whisker-like fibers for bone tissue engineering. *Mater. Sci. Eng. C Mater. Biol. Appl.* **2019**, *100*, 341–353. [CrossRef] [PubMed]
81. Aguilar, A.; Zein, N.; Harmouch, E.; Hafdi, B.; Bornert, F.; Offner, D.; Clauss, F.; Fioretti, F.; Huck, O.; Benkirane-Jessel, N.; et al. Application of chitosan in bone and dental engineering. *Molecules* **2019**, *24*, 3009. [CrossRef] [PubMed]
82. Bi, Y.G.; Lin, Z.T.; Deng, S.T. Fabrication and characterization of hydroxyapatite/sodium alginate/chitosan composite microspheres for drug delivery and bone tissue engineering. *Mater. Sci. Eng. C Mater. Biol. Appl.* **2019**, *100*, 576–583. [CrossRef] [PubMed]
83. Padalhin, A.R.; Lee, B.T. Hemostasis and bone regeneration using chitosan/gelatin-BCP bi-layer composite material. *ASAIO J.* **2019**, *65*, 620–627. [CrossRef] [PubMed]
84. Kjalarsdottir, L.; Dyrfjord, A.; Dagbjartsson, A.; Laxdal, E.H.; Orlygsson, G.; Gislason, J.; Einarsson, J.M.; Ng, C.H.; Jonsson, H. Bone remodeling effect of a chitosan and calcium phosphate-based composite. *Regen. Biomater.* **2019**, *6*, 241–247. [CrossRef] [PubMed]
85. Ye, H.; Zhu, J.; Deng, D.; Jin, S.; Li, J.; Man, Y. Enhanced osteogenesis and angiogenesis by PCL/chitosan/Sr-doped calcium phosphate electrospun nanocomposite membrane for guided bone regeneration. *J. Biomater. Sci. Polym. Ed.* **2019**, *30*, 1505–1522. [CrossRef] [PubMed]
86. Zeng, J.; Xiong, S.; Ding, L.; Zhou, J.; Li, J.; Qiu, P.; Liao, X.; Xiong, L.; Long, Z.; Liu, S. Study of bone repair mediated by recombination BMP-2/ recombination CXC chemokine Ligand-13-loaded hollow hydroxyapatite microspheres/chitosan composite. *Life Sci.* **2019**, *234*, 116743. [CrossRef] [PubMed]
87. He, K.; Wang, X.; Kumta, S.; Qin, L.; Yan, Y.; Zhang, R.; Wang, X. Fabrication of a two-level tumor bone repair biomaterial based on a rapid prototyping technique. *Biofabrication* **2009**, *1*, 025003.
88. Liao, I.C.; Moutos, F.T.; Estes, B.T.; Zhao, X.; Guilak, F. Composite three-dimensional woven scaffolds with interpenetrating network hydrogels to create functional synthetic articular cartilage. *Adv. Funct. Mater.* **2013**, *23*, 5833–5839. [CrossRef] [PubMed]

89. Nettles, D.L.; Elder, S.H.; Gilbert, J.A. Potential use of chitosan as a cell scaffold material for cartilage tissue engineering. *Tissue Eng.* **2002**, *8*, 1009–1016. [CrossRef] [PubMed]
90. Freedman, B.R.; Mooney, D.J. Biomaterials to Mimic and Heal Connective Tissues. *Adv. Mater.* **2019**, *31*, e1806695. [CrossRef] [PubMed]
91. Jin, R.; Moreira Teixeira, L.S.; Dijkstra, P.J.; Karperien, M.; van Blitterswijk, C.A.; Zhong, Z.Y.; Feijen, J. Injectable chitosan-based hydrogels for cartilage tissue engineering. *Biomaterials* **2009**, *30*, 2544–2551.
92. Liu, M.; Zeng, X.; Ma, C.; Yi, H.; Ali, Z.; Mou, X.; Li, S.; Deng, Y.; He, N. Injectable hydrogels for cartilage and bone tissue engineering. *Bone Res.* **2017**, *5*, 17014. [CrossRef] [PubMed]
93. Huang, H.; Zhang, X.; Hu, X.; Dai, L.; Zhu, J.; Man, Z.; Chen, H.; Zhou, C.; Ao, Y. Directing chondrogenic differentiation of mesenchymal stem cells with a solid-supported chitosan thermogel for cartilage tissue engineering. *Biomed. Mater.* **2014**, *9*, 035008. [CrossRef] [PubMed]
94. Kuo, C.Y.; Chen, C.H.; Hsiao, C.Y.; Chen, J.P. Incorporation of chitosan in biomimetic gelatin/chondroitin-6-sulfate/hyaluronan cryogel for cartilage tissue engineering. *Carbohydr. Polym.* **2015**, *117*, 722–730. [CrossRef] [PubMed]
95. VandeVord, P.J.; Matthew, H.W.; DeSilva, S.P.; Mayton, L.; Wu, B.; Wooley, P.H. Evaluation of the biocompatibility of a chitosan scaffold in mice. *J. Biomed. Mater. Res.* **2002**, *59*, 585–590. [CrossRef] [PubMed]
96. Choi, B.; Kim, S.; Lin, B.; Wu, B.M.; Lee, M. Cartilaginous extracellular matrix-modified chitosan hydrogels for cartilage tissue engineering. *ACS Appl. Mater. Interfaces* **2014**, *6*, 20110–20121. [CrossRef] [PubMed]
97. Oprenyeszk, F.; Sanchez, C.; Dubuc, J.E.; Maquet, V.; Henrist, C.; Compere, P.; Henrotin, Y. Chitosan enriched three-dimensional matrix reduces inflammatory and catabolic mediators production by human chondrocytes. *PLoS ONE* **2015**, *10*, e0128362. [CrossRef] [PubMed]
98. Kong, Y.; Zhang, Y.; Zhao, X.; Wang, G.; Liu, Q. Carboxymethyl-chitosan attenuates inducible nitric oxide synthase and promotes interleukin-10 production in rat chondrocytes. *Exp. Ther. Med.* **2017**, *14*, 5641–5646. [CrossRef] [PubMed]
99. Huang, S.; Song, X.; Li, T.; Xiao, J.; Chen, Y.; Gong, X.; Zeng, W.; Yang, L.; Chen, C. Pellet coculture of osteoarthritic chondrocytes and infrapatellar fat pad-derived mesenchymal stem cells with chitosan/hyaluronic acid nanoparticles promotes chondrogenic differentiation. *Stem Cell Res. Ther.* **2017**, *8*, 264. [CrossRef] [PubMed]
100. Sionkowska, A.; Wisniewski, M.; Skopinska, J.; Kennedy, C.J.; Wess, T.J. Molecular interactions in collagen and chitosan blends. *Biomaterials* **2004**, *25*, 795–801. [CrossRef]
101. Haaparanta, A.M.; Jarvinen, E.; Cengiz, I.F.; Ella, V.; Kokkonen, H.T.; Kiviranta, I.; Kellomaki, M. Preparation and characterization of collagen/PLA, chitosan/PLA, and collagen/chitosan/PLA hybrid scaffolds for cartilage tissue engineering. *J. Mater. Sci. Mater. Med.* **2014**, *25*, 1129–1136. [CrossRef] [PubMed]
102. Su, J.Y.; Chen, S.H.; Chen, Y.P.; Chen, W.C. Evaluation of magnetic nanoparticle-labeled chondrocytes cultivated on a type II collagen-chitosan/poly(lactic-co-glycolic) acid biphasic scaffold. *Int. J. Mol. Sci.* **2017**, *18*, 87. [CrossRef] [PubMed]
103. Bhardwaj, N.; Nguyen, Q.T.; Chen, A.C.; Kaplan, D.L.; Sah, R.L.; Kundu, S.C. Potential of 3-D tissue constructs engineered from bovine chondrocytes/silk fibroin-chitosan for in vitro cartilage tissue engineering. *Biomaterials* **2011**, *32*, 5773–5781. [CrossRef] [PubMed]
104. Liu, J.; Fang, Q.; Yu, X.; Wan, Y.; Xiao, B. Chitosan-based nanofibrous membrane unit with gradient compositional and structural features for mimicking calcified layer in osteochondral matrix. *Int. J. Mol. Sci.* **2018**, *19*, 2330. [CrossRef] [PubMed]
105. Ye, K.; Felimban, R.; Traianedes, K.; Moulton, S.E.; Wallace, G.G.; Chung, J.; Quigley, A.; Choong, P.F.; Myers, D.E. Chondrogenesis of infrapatellar fat pad derived adipose stem cells in 3D printed chitosan scaffold. *PLoS ONE* **2014**, *9*, e99410. [CrossRef] [PubMed]
106. Alves, N.M.; Mano, J.F. Chitosan derivatives obtained by chemical modifications for biomedical and environmental applications. *Int. J. Biol. Macromol.* **2008**, *43*, 401–414. [CrossRef] [PubMed]
107. Freier, T.; Montenegro, R.; Shan Koh, H.; Shoichet, M.S. Chitin-based tubes for tissue engineering in the nervous system. *Biomaterials* **2005**, *26*, 4624–4632. [CrossRef] [PubMed]
108. Jayakumar, R.; Nwe, N.; Tokura, S.; Tamura, H. Sulfated chitin and chitosan as novel biomaterials. *Int. J. Biol. Macromol.* **2007**, *40*, 175–181. [CrossRef] [PubMed]

109. He, Q.; Zhang, T.; Yang, Y.; Ding, F. In vitro biocompatibility of chitosan-based materials to primary culture of hippocampal neurons. *J. Mater. Sci. Mater. Med.* **2009**, *20*, 1457–1466. [CrossRef] [PubMed]
110. Yuan, Y.; Zhang, P.; Yang, Y.; Wang, X.; Gu, X. The interaction of Schwann cells with chitosan membranes and fibers in vitro. *Biomaterials* **2004**, *25*, 4273–4278. [CrossRef] [PubMed]
111. Wrobel, S.; Serra, S.C.; Ribeiro-Samy, S.; Sousa, N.; Heimann, C.; Barwig, C.; Grothe, C.; Salgado, A.J.; Haastert-Talini, K. In vitro evaluation of cell-seeded chitosan films for peripheral nerve tissue engineering. *Tissue Eng. Part A* **2014**, *20*, 2339–2349. [CrossRef] [PubMed]
112. Gonzalez-Perez, F.; Cobianchi, S.; Geuna, S.; Barwig, C.; Freier, T.; Udina, E.; Navarro, X. Tubulization with chitosan guides for the repair of long gap peripheral nerve injury in the rat. *Microsurgery* **2015**, *35*, 300–308. [CrossRef] [PubMed]
113. Meyer, C.; Stenberg, L.; Gonzalez-Perez, F.; Wrobel, S.; Ronchi, G.; Udina, E.; Suganuma, S.; Geuna, S.; Navarro, X.; Dahlin, L.B.; et al. Chitosan-film enhanced chitosan nerve guides for long-distance regeneration of peripheral nerves. *Biomaterials* **2016**, *76*, 33–51. [CrossRef] [PubMed]
114. Stenberg, L.; Kodama, A.; Lindwall-Blom, C.; Dahlin, L.B. Nerve regeneration in chitosan conduits and in autologous nerve grafts in healthy and in type 2 diabetic Goto-Kakizaki rats. *Eur. J. Neurosci.* **2016**, *43*, 463–473. [CrossRef] [PubMed]
115. Stenberg, L.; Stossel, M.; Ronchi, G.; Geuna, S.; Yin, Y.; Mommert, S.; Martensson, L.; Metzen, J.; Grothe, C.; Dahlin, L.B.; et al. Regeneration of long-distance peripheral nerve defects after delayed reconstruction in healthy and diabetic rats is supported by immunomodulatory chitosan nerve guides. *BMC Neurosci.* **2017**, *18*, 53. [CrossRef] [PubMed]
116. Wang, H.; Zhao, Q.; Zhao, W.; Liu, Q.; Gu, X.; Yang, Y. Repairing rat sciatic nerve injury by a nerve-growth-factor-loaded, chitosan-based nerve conduit. *Biotechnol. Appl. Biochem.* **2012**, *59*, 388–394. [CrossRef] [PubMed]
117. Wang, X.; Hu, W.; Cao, Y.; Yao, J.; Wu, J.; Gu, X. Dog sciatic nerve regeneration across a 30-mm defect bridged by a chitosan/PGA artificial nerve graft. *Brain* **2005**, *128*, 1897–1910. [CrossRef] [PubMed]
118. Jiao, H.; Yao, J.; Yang, Y.; Chen, X.; Lin, W.; Li, Y.; Gu, X.; Wang, X. Chitosan/polyglycolic acid nerve grafts for axon regeneration from prolonged axotomized neurons to chronically denervated segments. *Biomaterials* **2009**, *30*, 5004–5018. [CrossRef] [PubMed]
119. Ding, F.; Wu, J.; Yang, Y.; Hu, W.; Zhu, Q.; Tang, X.; Liu, J.; Gu, X. Use of tissue-engineered nerve grafts consisting of a chitosan/poly(lactic-co-glycolic acid)-based scaffold included with bone marrow mesenchymal cells for bridging 50-mm dog sciatic nerve gaps. *Tissue Eng. Part A* **2010**, *16*, 3779–3790. [CrossRef] [PubMed]
120. Xue, C.; Hu, N.; Gu, Y.; Yang, Y.; Liu, Y.; Liu, J.; Ding, F.; Gu, X. Joint use of a chitosan/PLGA scaffold and MSCs to bridge an extra large gap in dog sciatic nerve. *Neurorehabil. Neural. Repair* **2012**, *26*, 96–106. [CrossRef] [PubMed]
121. Zhao, L.; Qu, W.; Wu, Y.; Ma, H.; Jiang, H. Dorsal root ganglion-derived Schwann cells combined with poly(lactic-co-glycolic acid)/chitosan conduits for the repair of sciatic nerve defects in rats. *Neural Regen. Res.* **2014**, *9*, 1961–1967. [CrossRef] [PubMed]
122. Hao, C.; Wang, W.; Wang, S.; Zhang, L.; Guo, Y. An overview of the protective effects of chitosan and acetylated chitosan oligosaccharides against neuronal disorders. *Mar. Drugs* **2017**, *15*, 89. [CrossRef] [PubMed]
123. Xu, Y.; Zhang, Q.; Yu, S.; Yang, Y.; Ding, F. The protective effects of chitooligosaccharides against glucose deprivation-induced cell apoptosis in cultured cortical neurons through activation of PI3K/Akt and MEK/ERK1/2 pathways. *Brain Res.* **2011**, *1375*, 49–58. [CrossRef] [PubMed]
124. Jiang, M.; Cheng, Q.; Su, W.; Wang, C.; Yang, Y.; Cao, Z.; Ding, F. The beneficial effect of chitooligosaccharides on cell behavior and function of primary Schwann cells is accompanied by up-regulation of adhesion proteins and neurotrophins. *Neurochem. Res.* **2014**, *39*, 2047–2457. [CrossRef] [PubMed]
125. Gong, Y.; Gong, L.; Gu, X.; Ding, F. Chitooligosaccharides promote peripheral nerve regeneration in a rabbit common peroneal nerve crush injury model. *Microsurgery* **2009**, *29*, 650–656. [CrossRef] [PubMed]
126. Wang, Y.; Zhao, Y.; Sun, C.; Hu, W.; Zhao, J.; Li, G.; Zhang, L.; Liu, M.; Liu, Y.; Ding, F.; et al. Chitosan degradation products promote nerve regeneration by stimulating schwann cell proliferation via miR-27a/FOXO1 Axis. *Mol. Neurobiol.* **2016**, *53*, 28–39. [CrossRef] [PubMed]

127. Zhao, Y.; Wang, Y.; Gong, J.; Yang, L.; Niu, C.; Ni, X.; Wang, Y.; Peng, S.; Gu, X.; Sun, C.; et al. Chitosan degradation products facilitate peripheral nerve regeneration by improving macrophage-constructed microenvironments. *Biomaterials* **2017**, *134*, 64–77. [CrossRef] [PubMed]
128. Cui, T.; Yan, Y.; Zhang, R.; Liu, L.; Xu, W.; Wang, X. Rapid prototyping of a double layer polyurethane-collagen conduit for peripheral nerve regeneration. *Tissue Eng. C* **2008**, *15*, 1–9. [CrossRef] [PubMed]
129. Wang, X.; Cui, T.; Yan, Y.; Zhang, R. Peroneal nerve regeneration along a new polyurethane-collagen guide conduit. *J. Bioact. Compat. Polym.* **2009**, *24*, 109–127. [CrossRef]
130. He, K.; Wang, X. Rapid prototyping of tubular polyurethane and cell/hydrogel construct. *J. Bioact. Compat. Polym.* **2011**, *26*, 363–374.
131. Wang, X.; Rijff, B.L.; Khang, G. A building block approach into 3D printing a multi-channel organ regenerative scaffold. *J. Stem Cell Res. Ther.* **2015**, *11*, 1403–1411.
132. Kim, I.Y.; Seo, S.J.; Moon, H.S.; Yoo, M.K.; Park, I.Y.; Kim, B.C.; Cho, C.S. Chitosan and its derivatives for tissue engineering applications. *Biotechnol. Adv.* **2008**, *26*, 1–21. [CrossRef] [PubMed]
133. Hoekstra, R.; Chamuleau, R.A. Recent developments on human cell lines for the bioartificial liver. *Int. J. Artif. Organs* **2002**, *25*, 182–191. [CrossRef] [PubMed]
134. Kang, I.K.; Moon, J.S.; Jeon, H.M.; Meng, W.; Kim, Y.I.; Hwang, Y.J.; Kim, S. Morphology and metabolism of Ba-alginate encapsulated hepatocytes with galactosylated poly(allyl amine) and poly(vinyl alcohol) as extracellular matrices. *J. Mater. Sci. Mater. Med.* **2005**, *16*, 533–539. [CrossRef] [PubMed]
135. Li, J.; Pan, J.; Zhang, L.; Yu, Y. Culture of hepatocytes on fructose-modified chitosan scaffolds. *Biomaterials* **2003**, *24*, 2317–2322. [CrossRef]
136. Park, I.-K.; Yang, J.; Jeong, H.J.; Bom, H.S.; Harada, I.; Akaike, T.; Kim, S.I.; Cho, C.S. Galactosylated chitosan as a synthetic extracellular matrix for hepatocytes attachment. *Biomaterials* **2003**, *24*, 2331–2337. [CrossRef]
137. Chung, T.; Yang, J.; Akaike, T.; Cho, K.Y.; Nah, J.W.; Kim, S.I.; Cho, C.S. Preparation of alginate/galactosylated chitosan scaffold for hepatocyte attachment. *Biomaterials* **2002**, *23*, 2827–2834. [CrossRef]
138. Yu, Y.; Moncal, K.K.; Li, J.; Peng, W.; Rivero, I.; Martin, J.A.; Ozbolat, I.T. Three-dimensional bioprinting using self-assembling scalable scaffold-free "tissue strands" as a new bioink. *Sci. Rep.* **2016**, *6*, 28714. [CrossRef] [PubMed]
139. Li, J.; Pan, J.; Zhang, L.; Guo, X.; Yu, Y. Culture of primary rat hepatocytes within porous chitosan scaffolds. *J. Biomed. Mater. Res. A* **2003**, *67*, 938–943. [CrossRef] [PubMed]
140. Yan, Y.; Wang, X.; Xiong, Z.; Liu, H.; Liu, F.; Lin, F.; Wu, R.; Zhang, R.; Lu, Q. Direct construction of a three-dimensional structure with cells and hydrogel. *J. Bioact. Compat. Polym.* **2005**, *20*, 259–269. [CrossRef]
141. Li, S.; Yan, Y.; Xiong, Z.; Weng, C.; Zhang, R.; Wang, X. Gradient hydrogel construct based on an improved cell assembling system. *J. Bioact. Compat. Polym.* **2009**, *24*, 84–99. [CrossRef]
142. Wang, X.; Yan, Y.; Pan, Y.; Xiong, Z.; Liu, H.; Cheng, J.; Liu, F.; Lin, F.; Wu, R.; Zhang, R.; et al. Generation of three-dimensional hepatocyte/gelatin structures with rapid prototyping system. *Tissue Eng.* **2006**, *12*, 83–90. [CrossRef] [PubMed]
143. Xu, W.; Wang, X.; Yan, Y.; Zheng, W.; Xiong, Z.; Lin, F.; Wu, R.; Zhang, R. Rapid prototyping three-dimensional cell/gelatin/fibrinogen constructs for medical regeneration. *J. Bioact. Compat. Polym.* **2007**, *22*, 363–377. [CrossRef]
144. Zhang, T.; Yan, Y.; Wang, X.; Xiong, Z.; Lin, F.; Wu, R.; Zhang, R. Three-dimensional gelatin and gelatin/hyaluronan hydrogel structures for traumatic brain injury. *J. Bioact. Compat. Polym.* **2007**, *22*, 19–29. [CrossRef]
145. Xu, M.; Yan, Y.; Liu, H.; Yao, Y.; Wang, X. Control adipose-derived stromal cells differentiation into adipose and endothelial cells in a 3-D structure established by cell-assembly technique. *J. Bioact. Compat. Polym.* **2009**, *24*, 31–47. [CrossRef]
146. Xu, M.; Wang, X.; Yan, Y.; Yao, R.; Ge, Y. A cell-assembly derived physiological 3D model of the metabolic syndrome, based on adipose-derived stromal cells and a gelatin/alginate/fibrinogen matrix. *Biomaterials* **2010**, *31*, 3868–3877. [CrossRef] [PubMed]
147. Yao, R.; Zhang, R.; Wang, X. Design and evaluation of a cell microencapsulating device for cell assembly technology. *J. Bioact. Compat. Polym.* **2009**, *24*, 48–62. [CrossRef]
148. Yao, R.; Zhang, R.; Yan, Y.; Wang, X. In vitro angiogenesis of 3D tissue engineered adipose tissue. *J. Bioact. Compat. Polym.* **2009**, *24*, 5–24.

149. Xu, Y.; Li, D.; Wang, X. Liver manufacturing approaches: The thresholds of cell manipulation with bio-friendly materials for multifunctional organ regeneration. In *Organ Manufacturing*; Wang, X., Ed.; Nova Science Publishers Inc.: Hauppauge, NY, USA, 2015; pp. 201–225.
150. Wang, X. Overview on biocompatibilities of implantable biomaterials. In *Advances in Biomaterials Science and Biomedical Applications in Biomedicine*; Lazinica, R., Ed.; In Tech: Rijeka, Croatia, 2013; pp. 111–155.
151. Wang, X.; Tuomi, J.; Mäkitie, A.A.; Poloheimo, K.-S.; Partanen, J.; Yliperttula, M. The integrations of biomaterials and rapid prototyping techniques for intelligent manufacturing of complex organs. In *Advances in Biomaterials Science and Applications in Biomedicine*; Lazinica, R., Ed.; In Tech: Rijeka, Croatia, 2013; pp. 437–463.
152. Wang, X.; Yan, Y.; Zhang, R. Gelatin-based hydrogels for controlled cell assembly. In *Biomedical Applications of Hydrogels Handbook*; Ottenbrite, R.M., Ed.; Springer: New York, NY, USA, 2010; pp. 269–284.
153. Wang, X. Spatial effects of stem cell engagement in 3D printing constructs. *J. Stem Cells Res. Rev. Rep.* **2014**, *1*, 5–9.
154. Yan, Y.; Wang, X.; Pan, Y.; Liu, H.; Cheng, J.; Xiong, Z.; Lin, F.; Wu, R.; Zhang, R.; Lu, Q. Fabrication of viable tissue-engineered constructs with 3D cell-assembly technique. *Biomaterials* **2005**, *26*, 5864–5871. [CrossRef] [PubMed]
155. Li, S.; Xiong, Z.; Wang, X.; Yan, Y.; Liu, H.; Zhang, R. Direct fabrication of a hybrid cell/hydrogel construct by a double-nozzle assembling technology. *J. Bioact. Compat. Polym.* **2009**, *24*, 249–265.
156. Miguel, S.P.; Cabral, C.S.D.; Moreira, A.F.; Correia, I.J. Production and characterization of a novel asymmetric 3D printed construct aimed for skin tissue regeneration. *Colloids Surf. B Biointerfaces* **2019**, *181*, 994–1003. [CrossRef] [PubMed]
157. Kim, S.W.; Kim, D.Y.; Roh, H.H.; Kim, H.S.; Lee, J.W.; Lee, K.Y. Three-dimensional bioprinting of cell-laden constructs using polysaccharide-based self-healing hydrogels. *Biomacromolecules* **2019**, *20*, 1860–1866. [CrossRef] [PubMed]
158. Demirtas, T.T.; Irmak, G.; Gumusderelioglu, M. A bioprintable form of chitosan hydrogel for bone tissue engineering. *Biofabrication* **2017**, *9*, 035003. [CrossRef] [PubMed]
159. Xi, W.; Kong, F.; Yeo, J.C.; Yu, L.; Sonam, S.; Dao, M.; Gong, X.; Lim, C.T. Soft tubular microfluidics for 2D and 3D applications. *Proc. Natl. Acad. Sci. USA* **2017**, *114*, 10590–10595. [CrossRef] [PubMed]
160. Wang, J. Development of a Combined 3D Printer and Its Application in Complex Organ Construction. Master's Thesis, Tsinghua University, Beijing, China, 2014.
161. Wang, X.; He, K.; Zhang, W. Optimizing the fabrication processes for manufacturing a hybrid hierarchical polyurethane-cell/hydrogel construct. *J. Bioact. Compat. Polym.* **2013**, *28*, 303–319. [CrossRef]
162. Huang, Y.; He, K.; Wang, X. Rapid Prototyping of a hybrid hierarchical polyurethane-cell/hydrogel construct for regenerative medicine. *Mater. Sci. Eng. C* **2013**, *33*, 3220–3229. [CrossRef] [PubMed]
163. Wang, X.; Liu, C. Fibrin hydrogels for endothelialized liver tissue engineering with a predesigned vascular network. *Polymers* **2018**, *10*, 1084. [CrossRef] [PubMed]
164. Wang, X. Bioartificial organ manufacturing technologies. *Cell Transplant.* **2018**, *27*, 5–17. [CrossRef] [PubMed]
165. Wang, X. 3D printing of tissue/organ analogues for regenerative medicine. In *Handbook of Intelligent Scaffolds for Regenerative Medicine*, 2nd ed.; Pan Stanford Publishing: Palo Alto, CA, USA, 2016; pp. 557–570.
166. Wang, X.; Ao, Q.; Tian, X.; Fan, J.; Tong, H.; Hou, W.; Bai, S. Gelatin-based hydrogels for organ 3D bioprinting. *Polymers* **2017**, *9*, 401. [CrossRef] [PubMed]
167. Wang, X.; Ao, Q.; Tian, X.; Fan, J.; Wei, Y.; Hou, W.; Tong, H.; Bai, S. 3D bioprinting technologies for hard tissue and organ engineering. *Materials* **2016**, *9*, 802. [CrossRef] [PubMed]
168. Xu, Y.; Wang, X. Fluid and cell behaviors along a 3D printed alginate/gelatin/fibrin channel. *Bioeng. Biotech.* **2015**, *112*, 1683–1695. [CrossRef] [PubMed]
169. Wang, X.; Wang, J. Vascularization and adipogenesis of a spindle hierarchical adipose-derived stem cell/collagen/ alginate-PLGA construct for breast manufacturing. *IJITEE* **2015**, *4*, 1–8.
170. Liu, F.; Liu, C.; Chen, Q.; Ao, Q.; Tian, X.; Fan, J.; Tong, H.; Wang, X. Progress in organ 3D bioprinting. *Int. J. Bioprint.* **2017**, *4*, 1–15. [CrossRef]
171. Wang, X. Editorial: Drug delivery design for regenerative medicine. *Curr. Pharm. Des.* **2015**, *21*, 1503–1505. [CrossRef] [PubMed]

172. Zhao, X.; Du, S.; Chai, L.; Xu, Y.; Liu, L.; Zhou, X.; Wang, J.; Zhang, W.; Liu, C.-H.; Wang, X. Anti-cancer drug screening based on an adipose-derived stem cell/hepatocyte 3D printing technique. *J. Stem Cell Res. Ther.* **2015**, *5*. [CrossRef]
173. Liu, L.; Zhou, X.; Xu, Y.; Zhang, W.M.; Liu, C.-H.; Wang, X.H. Controlled release of growth factors for regenerative medicine. *Curr. Pharm. Des.* **2015**, *21*, 1627–1632. [CrossRef] [PubMed]
174. Xu, Y.; Wang, X. 3D biomimetic models for drug delivery and regenerative medicine. *Curr. Pharm. Des.* **2015**, *21*, 1618–1626. [CrossRef] [PubMed]
175. Zhou, X.; Liu, C.; Zhao, X.; Wang, X. A 3D bioprinting liver tumor model for drug screening. *World J. Pharm. Pharm. Sci.* **2016**, *5*, 196–213. [CrossRef]

 micromachines

Review

The Applications of 3D Printing for Craniofacial Tissue Engineering

Owen Tao [1], Jacqueline Kort-Mascort [2], Yi Lin [3], Hieu M. Pham [1], André M. Charbonneau [1], Osama A. ElKashty [1,4], Joseph M. Kinsella [2] and Simon D. Tran [1,*]

1 McGill Craniofacial Tissue Engineering and Stem Cells Laboratory, Faculty of Dentistry, McGill University, 3640 University Street, Montreal, QC H3A 0C7, Canada
2 Department of Bioengineering, McGill University, 817 Sherbrook Street West, Montreal, QC H3A 0C3, Canada
3 Department of Orthodontics, Guanghua School of Stomatology, Hospital of Stomatology, Sun Yat-sen University, 56 Lingyuan Road West, Guangzhou 510055, China
4 Oral Pathology Department, Faculty of Dentistry, Mansoura University, Mansoura 22123, Egypt
* Correspondence: simon.tran@mcgill.ca

Received: 20 June 2019; Accepted: 11 July 2019; Published: 17 July 2019

Abstract: Three-dimensional (3D) printing is an emerging technology in the field of dentistry. It uses a layer-by-layer manufacturing technique to create scaffolds that can be used for dental tissue engineering applications. While several 3D printing methodologies exist, such as selective laser sintering or fused deposition modeling, this paper will review the applications of 3D printing for craniofacial tissue engineering; in particular for the periodontal complex, dental pulp, alveolar bone, and cartilage. For the periodontal complex, a 3D printed scaffold was attempted to treat a periodontal defect; for dental pulp, hydrogels were created that can support an odontoblastic cell line; for bone and cartilage, a polycaprolactone scaffold with microspheres induced the formation of multiphase fibrocartilaginous tissues. While the current research highlights the development and potential of 3D printing, more research is required to fully understand this technology and for its incorporation into the dental field.

Keywords: 3D printing; additive manufacturing; bioprinting; dentistry; oral and maxillofacial regions; tissue engineering

---

1. Introduction

Three-dimensional (3D) printing is an emerging additive manufacturing technique capable of building complex 3D geometric structures, which can be used as scaffolds for craniofacial tissue engineering including the fabrication of biocompatible polymeric implants, the replication of intricate matrix geometries, and the development of biodegradable scaffolds, to cultivate transplantable tissues or organ replacements.

3D printing technology started in 1990, and it was mainly focused on fabricating scaffolds constituted of synthetic inks. It was not until the last decade when the technique evolved to what we currently know as bioprinting [1,2]. The development of bioinks, biocompatible soft materials that contain biological components such as cells or naturally derived matrices, promoted tissue engineering applications [3,4].

Several studies have successfully developed structures with relevant characteristics for regenerative dentistry—hydroxyapatite (HA) modified hydrogels have been reported suitable for bone bioprinting due to their osteosupportive and osteoinductive properties [5]. Additionally, polymers such as polycaprolactone (PCL) have also been reinforced with HA particles and 3D printed, resulting in scaffolds with good bioactivity shown by their in vitro apatite-forming ability [6].

Current bioprinting techniques include inkjet, stereolithography, laser-induced forward transfer (LIFT), and extrusion [7]. All of them can encapsulate cells or particles of interest in the hydrogel bioink. After printing, all these techniques provide cells with a 3D environment that mimics the biological conditions found in vivo. These techniques allow the design and fast fabrication of multiple models with the same architecture and dimensions of the original design with variables that can be easily controlled and manipulated for experimental purposes.

3D printing is a novel technique that is rapidly evolving and can become an important tool for the development of tissue-like constructs for oral surgery or translational research of biology and disease in dentistry. In this review, we will examine the types of 3D printing methodologies and materials as well as the applications of 3D printing in specific fields of dentistry such as the periodontal tissues, dental pulp, bone, and cartilage.

2. Three-Dimensional Printing Methodologies

2.1. Inkjet Printing

The principle behind inkjet printing consists of introducing a small volume change upstream of the nozzle, creating a pressure change, which results in a droplet ejection downstream (Figure 1A). This is performed by an inkjet head system which can be either piezoelectric or thermal induced. Thermal induced heads use a resistor as their heating structure. When current passes through the resistor, the fluid in contact is vaporized, creating a bubble that expands in the reservoir. This increases the pressure causing a droplet ejection through the nozzle [8]. Piezoelectric heads cause the volumetric change by applying a voltage pulse to the piezoelectric material [9].

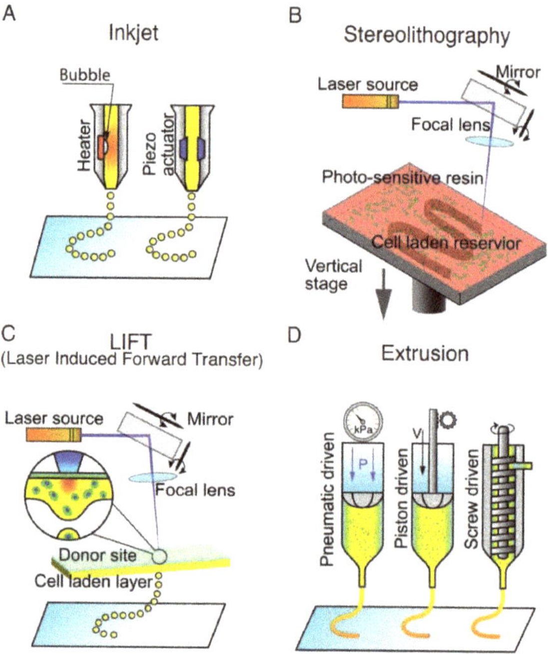

Figure 1. Schematic of various 3D printing methodologies. (**A**) Inkjet. A heater or piezo actuator deposits droplets. (**B**) Stereolithography. Layer by layer photopolymerization of a liquid resin by laser. (**C**) Laser induced forward transfer. Droplets of the material induced by a laser source. (**D**) Extrusion. Material exiting a nozzle that is pneumatic, piston, or screw driven. Reproduced with permission from [7].

A wide range of powder materials such as polymers, ceramics, proteins, and cells can be processed using this technique. However, the ink's viscosity is limited to 5–20 Pa.s to avoid high ejection pressures or continuous flow of the material. The main advantages of this technique include high-speed printability, low cost, and the possibility to encapsulate cells in the material [10].

For bone regeneration, bioink patterns using inkjet printing have been studied to control osteoblast differentiation in vitro and bone formation in vivo. A study created patterns of bone morphogenetic protein-2 (BMP-2) within microporous scaffolds (made from DermaMatrix, containing various extracellular molecules such as collagen and fibronectin) and observed cell differentiation in vitro and tissue formation in vivo in the patterned areas [11]. Additionally, another study used inkjet printing to 3D print calcium phosphate scaffolds and incorporated a collagen coating. The implanted scaffolds were osteoconductive while being biodegradable [12].

2.2. Laser-Assisted 3D Printing

Although less commonly used, laser printing technology has emerged from LIFT technology as a promising method for tissue engineering. It has prominent advantages in terms of bioprinting and is also known as laser-assisted bioprinting (LAB). LIFT assisted printers or LAB basically have three main components: (1) a pulsed laser source, (2) a target serving as a support for the printing materials, usually a transparent glass slide or ribbon, and (3) a receiving substrate to collect the materials. During printing, a focused laser pulse stimulates a small area of the target, which comprises an energy-absorbing layer on the surface and bioink solution underneath. Then a portion of the energy-absorbing layer is evaporated, resulting in the formation of a droplet that is collected by the receiving substrate and crosslinked therein [13,14].

Unlike inkjet printers, laser-assisted printers are equipped with no nozzles, obviating direct contact between the dispenser and the bioinks and therefore minimizes the problem of materials/cells clogging. With that, they are compatible with more materials, especially those with high viscosity (1–300 mPa/s), and can maintain cell viability higher than 95% [15]. Those benefits along with products with higher resolution makes laser-assisted printing a promising technology for tissue engineering.

Variations of laser-assisted 3D printing include selective laser sintering (SLS), stereolithography (SLA) (Figure 1B) and LIFT (Figure 1C). In particular, SLS has been extensively used for the regeneration of tissue with complex anatomy like craniofacial bone or cartilage. Developed by Carl Deckard for his Master's thesis at the University of Texas in 1989, SLS uses a high-powered carbon dioxide laser to create structures by fusing the powder layer-by-layer with the underlying powder as support [16]. The laser beam fuses powders selectively on the basis of sectional data from computer-assisted design (CAD). After a layer is created, the powder bed descends and another layer is rolled over. Such process is repeated until the scaffold is completed [17]. SLS can produce tissue engineering scaffolds from a variety of powder materials, including metals, bio-ceramics, and synthetic polymers like polylactic acid (PLA), PCL, poly ethyl ether ketone (PEEK), and poly ether ketone ketone (PEKK) [18,19]. Some researchers also include HA powders in the polymers to increase the osteoinductivity of bone scaffolds [20]. Natural polymers cannot be utilized with this technique due to the high temperatures generated by the laser during printing. However, similar to growth factors, they can be incorporated into the scaffolds post-printing.

PCL is an advantageous material in SLS, mainly because of its low melting (59–64 °C) and glass-transition temperatures (−60 °C) that facilitate the prototyping process. SLS-printed PCL scaffolds have been used to repair the periodontal [21], craniofacial bone, or osteochondral defects [22–24], and were proved to be biocompatible with adequate strength. Another polymer that has been recently applied in craniofacial regeneration using the SLS technique is PEEK [19], which has more favorable mechanical properties for stress loading than PCL. The manufacturing strategy, however, is not much different from that of PCL scaffolds. Higher performance polymers like PEKK have also been successfully printed by SLS technique [25,26] and introduced in the application of the craniofacial area [27,28].

Nowadays, no polymer processing methods can compete with the SLS regarding the fabrication flexibility and complexity of the 3D shapes obtained. However, the advancement of laser-assisted technologies is obviously restricted by the complicated control of the laser printing system and concerns about the side effect of laser exposure.

2.3. Extrusion

Fused deposition modeling (FDM) is the most common 3D printing technique. It uses a continuous filament made of thermoplastic polymer that is melted at the nozzle into a semi-liquid state and then extruded on a platform or on top of the previous layer (Figure 1D). The material fuses together to create a continuous structure after solidifying at room temperature. The quality of the extruded filament may be modified by adjusting the printing velocity, layer thickness, and printing orientation [29]. Some of the most common materials used for this technique are polycarbonate (PC), acrylonitrile butadiene styrene (ABS), PCL, and PLA due to their low melting points compared to other thermoplastics [30].

The main advantages of this technique are the affordability, high-speed printing, and the potential of printing multiple materials at the same time when working with a multi-nozzle printer [29]. Some disadvantages include the limitation to use only thermoplastic materials and the inability to embed cells in the material since thermoplastics melt at temperatures higher than 37 °C.

FDM-printed coated scaffolds and composite materials have been proven as useful tools for bone tissue engineering and regeneration. PCL scaffolds with freeze-dried platelet-rich plasma (PRP) were implanted in rats, and the results show that it can promote osteogenic differentiation of dental pulp stem cells and induce bone formation [31]. Additionally, anatomically shaped molar scaffolds made of PCL and hydroxyapatite with 200-µm-diameter interconnecting microchannels were implanted in rats and growth factors (stromal-derived factor-1 (SDF1) and bone morphogenetic protein-7 (BMP7)) were perfused. This setup recruited more endogenous cells and generated more angiogenesis than the control group [32].

Three-dimensional plotting (3DP) is a technique very similar to FDM. It consists of extruding a viscous material from a cartridge using pneumatic or mechanical pressure through a nozzle onto a defined position in a platform [30]. Multiple cartridges are mounted in an XYZ stage and the position of each cartridge, the pressure, and temperature are controlled by a computer. As with FDM, this technique also allows the printing of heterogeneous structures with different materials. Printing and curing of the materials is also possible by extruding the reactive components using mixing nozzles, exposing each layer to UV light or heating the stage to stabilize the material after printing [33]. The material flexibility is the main advantage of this technique. Hydrogels, plastics, pastes, and solutions can be printed using this technique, and several of these can be biocompatible allowing cell encapsulation before printing. Some disadvantages of this method compared with FDM are the resolution and speed.

Multiple bioinks suitable for this technique have been proposed to promote bone tissue regeneration. Studies using 3D printed periodontal cells encapsulated in a bioink constituted of different ratios of gelatin methacrylate (GelMA) and poly (ethylene glycol) (PEG) dimethacrylate have proven useful to study periodontal ligament stem cell response to extracellular matrix components [34]. Additionally, polymer solutions based on methacrylated gelatin and methacrylated hyaluronic acid modified with HA particles were used to encapsulate human adipose-derived stem cells and bioprint structures, which proved to be a suitable material for bone bioprinting applications [5]. A summary of the 3D printing types and their potential applications can be found in Table 1.

Table 1. Summary of 3D printing types.

Type	Methodology	Applications
Inkjet	Pressure change upstream of nozzle resulting in a downstream droplet ejection.	Regenerative approach—Printing of complex ceramic-like structures to support guided tissue regeneration. Replacement approach—Drop-by-drop bioprinting of live cells for the cell aggregate approach.
Laser-Assisted	Laser pulse stimulates a small area of the target.	Regenerative approach—Creation of more complex scaffolds for guided tissue regeneration.
Extrusion	Material fuses together at room temperature after leaving the nozzle.	Regenerative approach—Can be used with many materials for the creation of simple biocompatible and biodegradable scaffolds for guided tissue regeneration.

3. Materials for Three-Dimensional Printing

3.1. Polymers

Polymer materials are composed of chemical compounds typically formed from carbon, hydrogen, oxygen, and nitrogen. These monomer structures are repeated and bound with itself to create a longer molecular chain [35]. Polymers have become the most popular choice for 3D bioprinting in biomedical applications as it is often inexpensive, biocompatible, biodegradable, and can easily be manipulated with regard to its mechanical, chemical, and biological properties [36,37]. The polymer's manipulative property is particularly important in 3D printing because the printability of a material is dependent on its viscosity [37]. The ink being printed should be stiff enough to support subsequent layers, however if it becomes too viscous, it may lead to blockage of the printing nozzle [37]. Blockage is further avoided by printing the material in its pre-polymerized form. However, a limitation to using synthetic polymers is that the printing process for synthetic polymers induces high temperatures in which cells and growth factors cannot survive or remain active. Thus, cells or biologically active components are incorporated post-extrusion [37].

Polymeric materials can be printed in various forms as well, including powder, filament, and sheet form [38]. The polymer category encompasses a wide variety of materials that can range from being soft to hard, or synthetic to natural, however, the most commonly used polymers in craniofacial tissue engineering include PCL, PEEK, PLA, poly(lactic-co-glycolic acid) (PLGA), and chitosan [2,39,40]; the choice of polymer will depend on the goal of the researcher. For example, PLA or PCL are popular choices for drug delivery purposes, while alginates and gelatin are more popular choices for cell encapsulation [37,38]. The wide variety in polymeric materials makes them highly versatile, and thus may be highly useful in dental tissue regeneration when being combined with the superior spatial resolution provided by 3D printing.

3.2. Ceramics

Ceramic materials consist of metals with inorganic calcium or phosphate salts (such as calcium silicate or β-tricalcium phosphate) and are generally osteoconductive and osteoinductive [41]. The composition of these scaffolds also allows them to last longer than hydrogels, permitting more time for structural support and for guided tissue regeneration. Although the properties of ceramic materials allow for cells to quickly proliferate and differentiate on the scaffold, a limitation lies in its inherent brittleness and poor mechanical strength—characteristics that may be necessary when dealing with load-bearing defects [41]. Several studies have however aimed to improve the effectiveness of ceramic materials by changing the pore size as well as through the addition of polymers, such as PCL or PLA. These alterations to the ceramic-based materials allow them to resemble better the mechanical properties of natural bone while being able to promote vascularization [42].

In a review by Jammalamadaka and Tappa, various 3D printing methodologies (such as extrusion [43–45], inkjet [12,46,47], and laser sintering [48,49]) for ceramic-based materials are mentioned, in addition to the use of sintering and freeze-drying methods post-printing to improve the mechanical properties of the scaffolds [42]. The FDM printing of ceramics is briefly outlined in a review by Obregon and colleagues, where scaffold manufacturing consists of three phases that use organic particles to facilitate flowability, which are then burned out with high temperatures leaving behind primarily the inorganic ceramic particles [37].

3.3. Composites

Printable composite materials are composed of a minimum of two different materials; mixtures for printable composites being used in dentistry are typically composed of copolymers, polymer-polymer mixtures, or polymer-ceramic mixtures; be ceramic-based or hydrogel-based; and can include the addition of biomolecules, carbon nanotubes, and metals [37,50,51]. The mixture will be dependent on the goal of the composite, but it is typically created to manipulate ink properties such as processability, printability, stiffness, and bioactivity [51]. By combining multiple materials, composite materials can harness the benefits of each individual material [50]. For example, the polymer PLA alone has great chemical and physical properties, however, may not be optimally biocompatible as it releases acidic compounds over time. Researchers overcame this issue by creating a composite containing PLA and ceramics such as calcium phosphate, which ultimately lessens the formation of acidic environments formed by PLA [51]. Composites can also be enhanced with silicate fillers and nanoparticles, which alter its viscosity and stiffness and ability to influence cell morphology [37].

Composite materials are frequently used in craniofacial regeneration due to its unique properties. While hard polymers exist, composite materials are more capable of mirroring complex tissues that withstand higher mechanical stress and loads such as bones and teeth, and thus are more favorable for craniofacial regeneration [37]. For example, in bone regeneration, researchers again have combined PLA with ceramic to take advantage of PLA's mechanical properties while overcoming its brittle nature [51]. Other uses for printable composites include cartilage regeneration and whole-tooth regeneration [37].

3.4. Cell Aggregates

Bioprinting of cell aggregates has been used as a scaffold-free methodology of creating tissue engineered constructs. These cell aggregates consist of spheroid structures, which can then be specifically positioned, creating for instance tubular or ring-like structures [52,53]. Although the constructs are primarily scaffold-free, the cells are usually encapsulated with a hydrogel material that is biocompatible and biodegradable, for cell survival and for mechanical support of the cell construct. The hydrogel also helps to prevent tissue fusion while the cells are maintained in the suspension reservoir of the 3D printer [54]. The use of cross-linking solutions, such as those containing $CaCl_2$ or gelatin, can help to further minimize cell aggregation [55,56]. As the pH of the bioink is important for cellular survival and scaffold integrity, a study by Lozano and colleagues used the addition of NaOH to stabilize the pH of a modified bio-polymer hydrogel [57].

The advantages of using scaffold-free constructs for tissue engineering include the absence of potentially toxic or immunogenic scaffold materials, as well as the ability to create high cell density constructs [37,58]. Limitations of the cell aggregate approach include the relatively time-consuming cellular fusion of the spheroids to create larger tissue structures (which may also create non-uniform structures) [54]. Certain advances have been made to minimize this limitation such as the development of multicellular cylinders as an alternative structure, which require up to four days to create the appropriate shape [53]. While most 3D printing of cell aggregate studies have been performed in vitro, there is a limited understanding of its potential in vivo and further studies must be performed to demonstrate its safety and feasibility as a scaffold-free construct for tissue engineering. A summary of the 3D printing materials and their potential applications can be found in Table 2.

Table 2. Summary of 3D printing materials for tissue engineering.

Type	Materials	Applications
Polymers	Compounds typically formed from carbon, hydrogen, oxygen, and nitrogen, such as PCL, PEEK, PLA, PLGA.	Regenerative approach—Uses biodegradable polymers as a guide for tissue regeneration.
Ceramics	Metals with inorganic calcium or phosphate salts (calcium silicate or β-tricalcium phosphate).	Regenerative approach—Longer-lasting ceramic-type scaffolds can permit more time for structural support and for guided tissue regeneration.
Composites	A combination of a minimum of two different materials, for instance copolymers, polymer-polymer mixtures, or polymer-ceramic mixtures.	Regenerative approach—Composites (such as PLA with ceramics) can be created to facilitate the regenerative approach by reducing the formation of acidic environments caused by PLA alone. Replacement approach—Composite hydrogels (such as those containing silica) can be created to facilitate the replacement approach by increasing gene expression of BMPs.
Cell Aggregates	Cell aggregates form spheroid structures, which are then used as a scaffold-free application of tissue regeneration.	Replacement approach—Post-printing fusion of spheroids create structures that can be used as replacements for damaged or missing tissues.

4. Pre-Clinical and Clinical Applications

4.1. Periodontal Complex

The concept of using 3D printing for periodontal tissue regeneration is to guide locally available cells to restore periodontal defects, termed guided tissue regeneration (GTR) [59]. These cells can use the support of a 3D printed scaffold, as well as surrounding growth factors, bioactive proteins, etc., to regenerate damaged periodontal tissues [60]. While epithelial tissues regenerate quickly, bone tissues require more time, and therefore a difficulty in periodontal regeneration lies in its tissue complexity (cementum, periodontal ligament, etc.) [59]. To address these challenges, the creation of multiphasic scaffolds has allowed for various properties within a scaffold, which mimics better the composition of the native periodontal complex [59].

PCL scaffolds have been 3D printed and used for periodontal tissue engineering. Improvements to these scaffolds have focused on several key aspects, as stated by Ivanovski and colleagues, relating to periodontal tissue engineering: (1) compartmentalized bone and periodontal attachment tissue formation; (2) cementum formation onto the root surface; (3) correctly oriented periodontal ligament fibers [61]. Lee and colleagues, for instance, 3D printed PCL-HA scaffolds with varying sizes of microchannels to create a compartmentalized multiphasic scaffold [62]. FDM was used to create these scaffolds, which had 100 μm microchannels designed for the cementum/dentin interface, 600 μm for the periodontal ligament (PDL), and 300 μm for the alveolar bone [62]. They found that in vivo implantation of dental pulp stem cells with the scaffold resulted in differentiation of the cell population into putative dentin/cementum, PDL, and alveolar bone [62]. Similarly, Li and colleagues used a freeze-dried PRP coating to improve the biological properties of PCL scaffolds [31]. This coating was applied to the 3D printed PCL scaffold for 5 min at room temperature, then placed at −80 °C for 30 min, followed by freeze-drying [31]. The freeze-dried PRP-PCL scaffolds induced significantly greater bone formation compared to traditional PRP-PCL or bare PCL scaffolds [31].

Bioprinting of PDL cells, creating a 3D hydrogel microarray, has been performed to screen for cell-biomaterial interactions [34]. The cells were bioprinted using a pressure-assisted valve-based bioprinting system placed within a sterile hood and controlled by a computer [34]. The pressure-based system replaces the need for any external stimulation, and thereby minimizes shear forces and high temperatures, allowing for the cells to survive the printing process [34]. This study has found that the viability of the printed periodontal cells was maintained throughout printing, and therefore this methodology can perhaps be used to 3D print periodontal cells directly into future scaffolds [34]. Likewise, Hamlet and colleagues examined alveolar bone regeneration through 3D hyaluronic acid hydrogels containing osteoblasts and found that their hydrogel provided a favorable environment and could stimulate osteogenic gene expression in vitro: they believed that this hydrogel could be optimized as a cell-delivering bioink for future 3D bioprinting applications [63].

The combination of cell sheets with 3D printed scaffolds has also been used for periodontal tissue regeneration. Vaquette and colleagues cultured PDL cell sheets in 24-well plates, which were then combined with an FDM printed PCL scaffold by folding the cell sheet over the scaffold (Figure 2A–C) [64]. They found that the scaffolds incorporating the cell sheet technology had better attachment onto a dentin surface than those without [64]. Farag and colleagues, however, aimed to improve the cell sheet technology by decellularizing the cell sheet after combination with the PCL scaffold (Figure 2D,E) [65]. The decellularization aimed to use the properties of the PDL extracellular matrix to promote periodontal regeneration, while minimizing the immunogenic effects of cellular material. They found that the decellularized cell sheet constructs upregulated the expression of mineralized tissue markers in PDL cells [65]. As an application of scaffold free bioprinting, Bakirci and colleagues developed a novel cell sheet based bioink for 3D bioprinting [66]. Cells were first grown on poly(N-isopropylacrylamide) coated surfaces, harvested, and centrifuged into cell sheet aggregates to be used for bioprinting applications [66]. Although this study developed a cell sheet based bioink using human skin fibroblasts, it illustrates the possibility in developing a PDL cell sheet based bioink, which could then be used for periodontal tissue engineering applications such as scaffolds in treating periodontal-related cases.

Clinically, Rasperini and colleagues reported the use of a SLS printed PCL scaffold to treat a periodontal defect (Figure 2F,G) [21]. A computed tomography scan of the patient's defect was taken to modify the scaffold design and to create a customized scaffold [21]. This scaffold consisted of an internal port for growth factor delivery and pegs perpendicular to the root to facilitate PDL formation [21]. At two weeks, the scaffold was removed and unfortunately the patient showed minimal evidence of bone repair [21]. However, this case highlighted a potential for the use of 3D printed scaffolds in treating periodontal-related cases.

4.2. Dental Pulp

The dental pulp is an unmineralized tissue beneath the mineralized hard exterior of the tooth, which plays a crucial role in tooth vitality; injury, alteration, and/or removal of the pulp may lead to tooth necrosis. Additionally, the pulp may also service to provide immunity, nutrition, and sensation as well [50]. As a result, there is a need to focus on protecting the pulp from trauma, and to regenerate the pulp should it be injured. While there has not been major success in pulp regeneration due to the challenges met in nurturing, revascularizing, and reinnervating the pulp tissue, a promising direction that researchers are exploring is the use of hydrogels to contain and nurture dental pulp cells [68]. By using hydrogels and other biomaterials as cellular scaffolds to mirror the native in vivo environment, researchers can promote cell growth, differentiation, and morphogenesis [50]. However, the major constraint with using hydrogels alone is that spatial manipulation is limited, i.e., researchers cannot fully control multicellular organization and interaction, thus the overall morphogenesis of the artificial gland or tissue. By applying the superior spatial control of 3D cell printing, this issue can be overcome in pulp tissue regeneration [69]. 3D cell printing technology would enable researchers to suspend and position various cells contained in hydrogels as they desire. For example, researchers could print odontoblastoid cells along the dentin walls while having fibroblasts towards the center of the pulp chamber [69]. Furthermore, the enhanced precision gained from 3D bioprinting would allow researchers to achieve specific cellular interactions, anisotropic mechanical properties, and desired distribution of growth factors [70]. While theoretically the use of 3D cell printing in pulp tissue regeneration sounds feasible, there is a lack of evidence for this to date. Several studies have shown the possibility of successfully 3D printing blood capillaries, however in vivo angiogenesis has not been exhibited in endodontics [37,69].

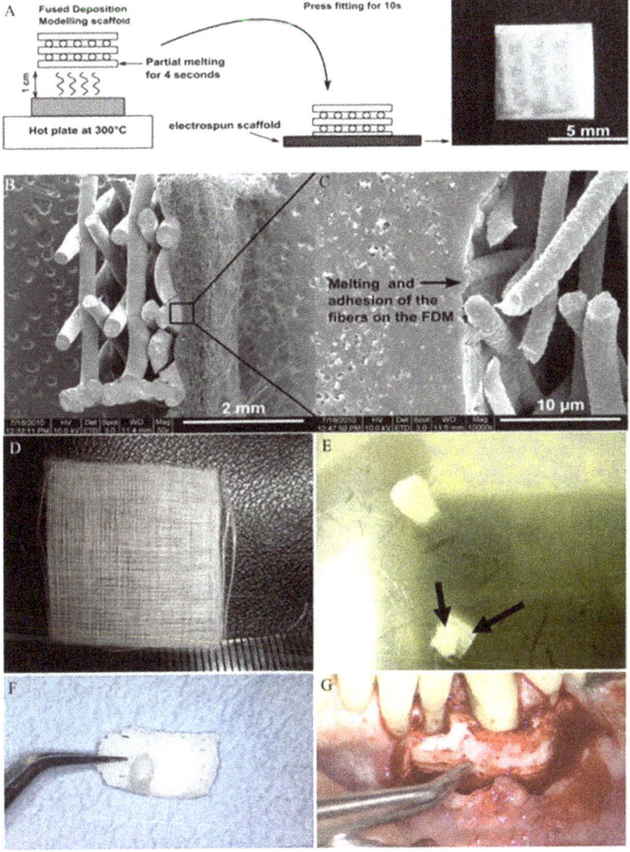

Figure 2. 3D Printed scaffolds for periodontal tissue engineering. (**A–C**) Schematic of the scaffold fabrication methodology (**A**). Cross-section showing the fusion of the electrospun fibers with the fused deposition modeling (FDM)-printed compartment of the scaffold (**B**,**C**). (**D**,**E**) Electron spun polycaprolactone (PCL) scaffold (**D**). The PCL scaffold attached to a decellularized sheet (**E**). (**F**,**G**) selective laser sintering (SLS)-printed PCL scaffold to be implanted in patient (**F**). Scaffold placement for implantation (**G**). Reproduced with permission from [21,65,67].

While there is a lack of in vivo studies to date, there are several studies that highlight the potential use of 3D cell printing in dental pulp regeneration. For example, in a study by Athirasala and colleagues, they showed that the mouse odontoblast-like cell line (OD21) could be supported in a novel hydrogel composed of alginate and dentin (Alg-Dent) [71]. While this study did not address more complicated experiments such as the possibility of vasculogenesis and/or angiogenesis, the use of human cells, and cell survivability in root canals and chamber, the study demonstrated the tunability and printability of the scaffold using 3D bioprinting technology. The success of this paper indicated the potential of the scaffold and how 3D bioprinting could further enhance the feasibility of this hydrogel in regenerative endodontics. Specifically, researchers would be able to localize growth factors and other nutrients precisely to the desired targets such as the peripheral dentin or central pulp to induce cell-specific regeneration as evident in this study.

As previously mentioned, 3D bioprinting methods allow researchers to achieve superior tunability, creating scaffolds that would not be possible without it. For example, in a recent study by Feng and colleagues, they compared two different techniques in fabricating a PLA scaffold, either molding via standard extrusion processes or 3D printed, and its influence on dental pulp cells. The results indicated that manufacturing techniques can influence differences in cell migration, morphology, and differentiation marker expression [72]. Another study by Hu and colleagues used 3D printed molds to create cellularized conduits for peripheral nerve regeneration, which showed comparable results to the use of autografts in repairing peripheral nerve defects (Figure 3) [73]. Thus, further studies should be explored, comparing current scaffold manufacturing methods and 3D printing and its effects on vasculogenesis and angiogenesis in support of pulp regeneration.

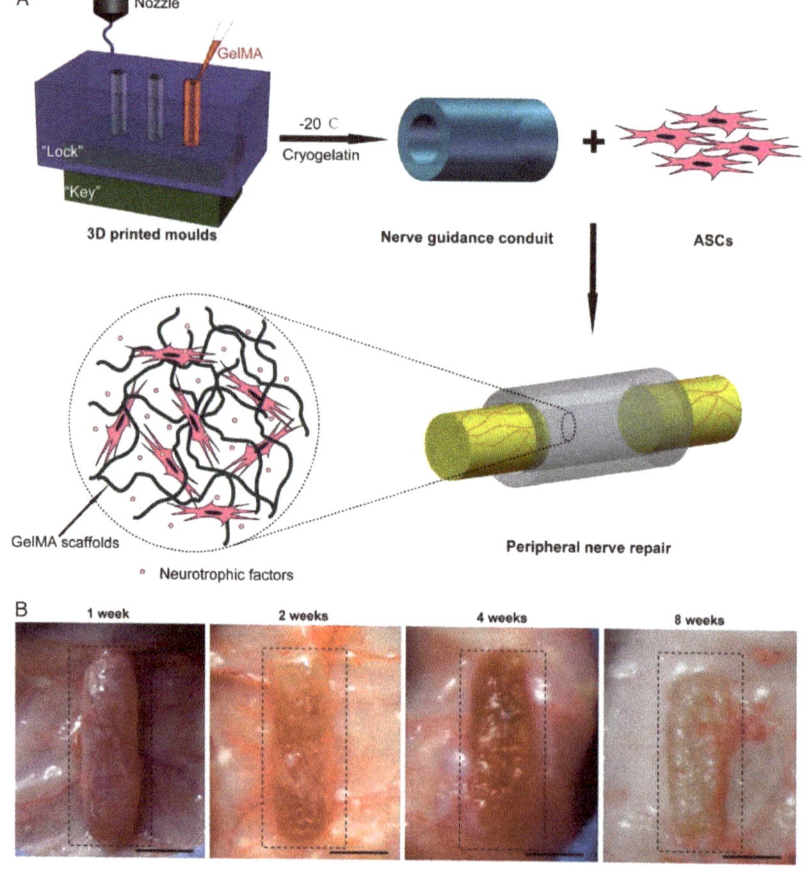

Figure 3. Cellularized conduits for peripheral nerve regeneration created using 3D printed molds. (**A**) Schematic of the conduit fabrication method. (**B**) Photographs of the rat dorsal side with the biodegradable nerve guidance conduit positioned subcutaneously. The figures are © 2016, Hu Y., Wu Y., et al. (https://doi.org/10.1038/srep32184) used under a Creative Commons Attribution 4.0 International License: http://creativecommons.org/licenses/by/4.0/.

Though there are not many studies to date highlighting the use of 3D printing in pulp regeneration, the current studies that demonstrate the superior tunability and modifications in the mechanical properties of currently viable scaffolds using 3D printing indicates the potential it may have in pulp regeneration. Further studies need to be explored by implementing the results, knowledge, and support gained from current studies in the possibility of inducing vasculogenesis, angiogenesis, and nutritional support in pulp tissue using 3D printing techniques.

4.3. *Cranio-Maxillofacial Tissues*

Craniofacial bones and cartilages comprise the craniofacial skeleton that impart specific appearance and function. It is challenging to reconstruct craniofacial structures due to the complex 3D geometry. With the combination of image-based extraction of craniofacial geometry and the ability to 3D-print shapes with high fidelity, 3D printing technologies are ideally suited for the manufacture of bone and cartilage scaffolds tailored to specific defects. The goal of craniofacial 3D reconstruction is to mimic the external and internal architecture of the host site and to provide essential framework for cell attachment and migration.

Bone is considered the second most transplanted tissue for defects due to trauma, osteoporosis, bone tumors, etc. Many types of biomaterials have been proposed to integrate the desirable properties, such as biocompatibility, printability, osteoconductivity, osteoinductivity, and mechanical properties, attempting to mimic the natural replacement of bone.

Bioceramics are the most commonly selected materials. They are usually composed of calcium and phosphate mineral phases, such as HA, β-tricalcium phosphate (TCP), or bioactive glasses (BGs). They exhibit outstanding biocompatibility and favorable biodegradability. Moreover, 3D printed ceramics can upregulate osteogenesis by creating a bioactive ion-rich cellular micro-environment and promote cell proliferation by close cell-cell interactions [74]. Even so, ceramic scaffolds are too brittle for implantation in load bearing craniofacial sites. Saijo and colleagues have confirmed this disadvantage by using HA/α-TCP composite scaffolds for maxillomandibular defects, which showed difficulties in composition and fabrication of an ideal scaffold to fulfill strength and dimensional requirements [75]. Still, Shao and colleagues have recently reported that ~10% Mg-substituted wollastonite had much higher flexural strength (31 MPa) than TCP and other calcium-silicate porous bioceramics [76]. By adding a range of metallic ions like Cu^{2+} and Co^{2+} into BGs, the angiogenic activity in vivo can be developed, which is beneficial for the healing process [77]. Compared to being used alone, they are more commonly incorporated with other biomaterials such as polymers for the enhancement of osteogenesis and osteoinductivity.

Polymers are another widely-used material that is superior in its printability and efficiency in promoting osteogenesis. The main concerns are its poor cellular interaction and low stiffness. PLA and PGA, for example, are now rarely used for bone scaffolds considering their low compressive strength and osteoconductivity. However, their co-polymer PLGA and another polyester, PCL, have remarkable osteoconductivity and better mechanical properties. By comparison, PCL has a lower rate of degradation and subsequently denser tissues generated [78]. Therefore, it is preferred to be used as the framework of composite scaffolds. As PCL is bioinert, other biological active components such as TCP, HA, decellularized trabecular bone, or growth factors were incorporated into the 3D printing system [21,79,80]. Furthermore, the acidic environment caused by the degradation products of PCL and its hydrophobic nature can be somewhat diminished by the inclusion of hydrophilic polymers like PEG and the surface coating of natural polymers like chitosan [81,82]. PCL scaffolds are well-suited for extrusion-based 3D-printing (Figure 4A), FDM for example, due to the relatively low melting points (62 °C). In recent decades, SLS has been developed as a more precise 3D printing technique that can fine-tune the porosity to optimize conditions for cell growth and proliferation. Additionally, a wide range of thermoplastic materials including high performance plastics with specific mechanical properties can be processed by SLS technique. The stiffness of PCL scaffolds manufactured by SLS has been reported to be ~15 to 300 MPa, values that are much higher than conventional 3D polymers but still lower than human trabecular bone within the condyle (120–450 MPa) or within the mandibular body (112–910 MPa) [22]. Metal-based scaffolds, on the other hand, are stiff enough but possess a considerably higher Young's modulus, which would lead to stress shielding issues and therefore failure of the implants. Polyaryletherketones (PEAKs) is a family of high-performance polymers with compatible Young's modulus to natural bone, which would be a suitable property for load bearing orthopedic and craniofacial implants. PEKK is by far a material with the most advantageous performance in the PEAKs family. The PEKK printed by SLS platforms have showed desirable mechanical properties, great biocompatibility, and osteointegration in a craniofacial bone defect model in vivo [18,27,28]. As a promising 3D printing material for bone tissue engineering, more evidence of success is needed for future applications.

The cartilaginous tissues in the craniofacial area primarily include the temporomandibular joint (TMJ) disc, the auricle cartilage, and the nasal cartilage. The bioinks used for cartilage reconstruction should be able to mimic the 3D architecture with mechanical anisotropic, nonlinear, and viscoelastic behavior analogous to native cartilage.

In early approaches, many hydrogels encapsulating the chondrocytes/mesenchymal stem cells (MSCs) with the capability of synthesizing extracellular matrix were fabricated by micro-extrusion technique. The cell-laden hydrogels ranged from natural polymers like alginate and collagen to synthetic polymers like gelatin metacrylamide (GelMA) and polyethylene glycol dimethacrylate (PEGDMA) [83–86]. To improve the mechanical properties, higher polymer concentration and viscosity were preferred. On the other hand, cells proliferate and differentiate towards cartilage tissue more readily within lower polymer concentrations. This dilemma increased the challenge of using the hydrogels alone to reconstruct the cartilaginous tissue. The most common solution is to include a stiffer thermoplastic polymer such as PCL to cell-laden hydrogels by coextrusion or other hybrid strategies (Figure 4B). PCL acts as a frame to reinforce the constructs, and by modulating the polymer percentage, the compressive equilibrium moduli in the range of articular cartilage can be achieved [87–89].

The degradation rate of PCL, which can be up to 2–3 years, is a potential limitation with such multi-material approaches, as residual filaments can act as a barrier to tissue formation. One alternative is applying polymers that have a higher rate of degradation, poly (hydroxymethylglycolide-co-caprolactone) (PHMGCL) or PLGA, for instance [91]. Yet the acidic by-products of PLGA that cause adverse inflammatory response still remains a concern for future applications. Tarafder and colleagues developed a region-variant TMJ disc scaffold by incorporating the specifically-aligned PCL with PLGA microspheres encapsulating TGFβ3 [92,93]. After seeding with MSCs, multiphase fibrocartilaginous tissues formed and significantly improved the healing process of the perforated disc. The dynamic function was also restored as no arthritis changes were observed on the condyle four weeks post-implantation.

Another way to reduce the residual PCL materials is to increase its porosity by the melt-electrowriting (MEW) technique, which is similar to FDM but using a nozzle tip equipped with voltage. PCL fabricated by MEW can be very thin, with a diameter down to 0.8 μm, and therefore the porosity can be high, up to 93%–98% [94]. In addition, the stiffness and yielding strains of the resultant scaffolds were within the range of native cartilage.

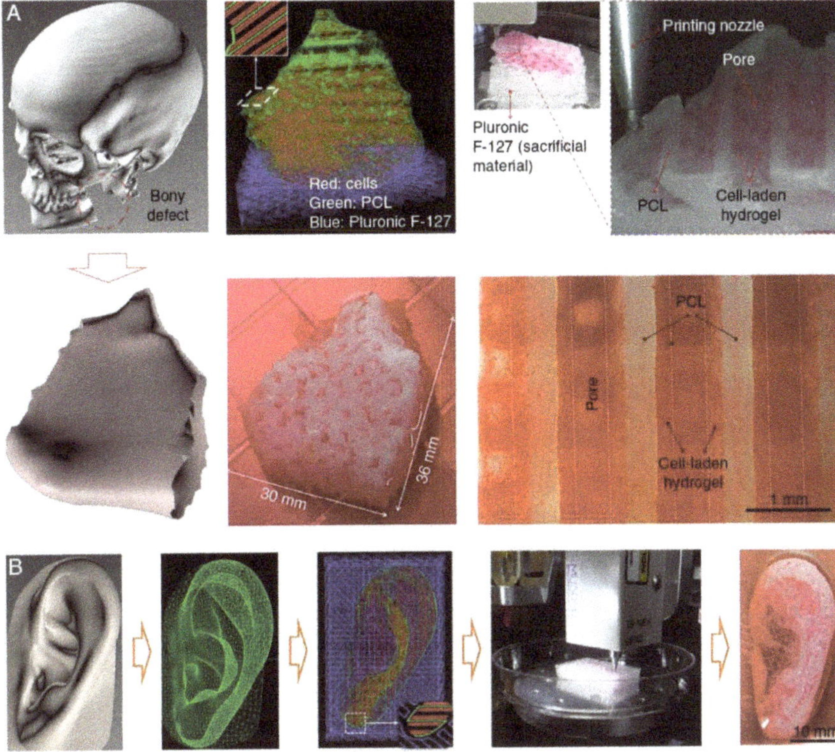

Figure 4. Craniofacial bone and cartilage reconstruction using PCL as a material for 3D printing. (**A**) Mandibular bone reconstruction. 3D defect model was obtained from the craniofacial CT image data followed by the design of dispensing paths of cells, PCL, and Pluronic F-127 with self-developed software. Multiple cartridges used to deliver and pattern the above ink materials were connected to a microscale nozzle, which dispensed the materials according to the design during 3D printing process. PCL was printed as the framework and the cell-laden hydrogel were dispensed to fill the pores, while Pluronic F-127 were used as sacrificing materials. The osteogenic potential of the scaffold was confirmed by Alizarin Red S staining after being cultured in osteogenic medium for 28 d. (**B**) Auricle cartilage construction. Similarly, a 3D computer-assisted (CAD) model of auricle can be developed from CT or MRI image data and generate a visualized motion program consisting of a command list for XYZ stage movements and air pressure actuation for 3D printing. The concentrations of different ingredients for 3D printing can be optimized by in vitro culture and related tests. Reproduced with permission from [90].

5. Conclusions

3D printing has the potential to revolutionize dentistry. This technique allows for a layer-by-layer construction of tissue engineering scaffolds—to create accurate, yet complex scaffold models for personalized patient treatments. Recently, there have been many advances in 3D printing for dentistry: for the periodontal complex, FDM printed scaffolds have been modified to induce greater bone formation, while an SLS printed scaffold has been applied for the first time in a human patient; for dental pulp, a 3D printed hydrogel could support odontoblast cell survivability; for bone and cartilage, modified bioceramic scaffolds induced greater angiogenesis and a modified PCL scaffold induced greater fibrocartilaginous tissue formation. While there are many in vitro studies examining the efficacy of 3D printed scaffolds for tissue engineering, further research must be performed to better understand the potential of these scaffolds in vivo, and to address any unprecedented safety concerns. As this technology develops, we expect to see a greater number of dental offices equipped with 3D printing technology, not only for the printing of crowns and dentures, but also for the purposes relating to tissue engineering.

Author Contributions: Conceptualization, O.T. and S.D.T.; investigation, O.T., J.K.-M., Y.L. and H.M.P.; writing—original draft preparation, O.T., J.K.-M., Y.L., H.M.P. and A.M.C.; writing—review and editing, O.A.E., J.M.K. and S.D.T.; supervision, S.D.T.; project administration, O.T. and S.D.T.

Funding: This research received no external funding.

Conflicts of Interest: The authors declare no conflict of interest.

References

1. Laurienzo, P.; Malinconico, M.; Motta, A.; Vicinanza, A. Synthesis and characterization of a novel alginate–poly(ethylene glycol) graft copolymer. *Carbohydr. Polym.* **2005**, *62*, 274–282. [CrossRef]
2. Dawood, A.; Marti, B.M.; Sauret-Jackson, V.; Darwood, A. 3D printing in dentistry. *Br. Dent. J.* **2015**, *219*, 521. [CrossRef] [PubMed]
3. Nichol, J.W.; Khademhosseini, A. Modular tissue engineering: Engineering biological tissues from the bottom up. *Soft Matter.* **2009**, *5*, 1312–1319. [CrossRef] [PubMed]
4. Derby, B. Printing and prototyping of tissues and scaffolds. *Science* **2012**, *338*, 921–926. [CrossRef] [PubMed]
5. Wenz, A.; Borchers, K.; Tovar, G.E.M.; Kluger, P.J. Bone matrix production in hydroxyapatite-modified hydrogels suitable for bone bioprinting. *Biofabrication* **2017**, *9*, 044103. [CrossRef] [PubMed]
6. Kim, J.W.; Shin, K.H.; Koh, Y.H.; Hah, M.J.; Moon, J.; Kim, H.E. Production of poly(ε-Caprolactone)/hydroxyapatite composite scaffolds with a tailored macro/micro-porous structure, high mechanical properties, and excellent bioactivity. *Materials* **2017**, *10*, 1123. [CrossRef] [PubMed]
7. Jiang, T.; Munguia-Lopez, J.G.; Flores-Torres, S.; Kort-Mascort, J.; Kinsella, J.M. Extrusion bioprinting of soft materials: An emerging technique for biological model fabrication. *Appl. Phys. Rev.* **2019**, *6*, 011310. [CrossRef]
8. Kumar, A.V.; Dutta, A.; Fay, J.E. Electrophotographic printing of part and binder powders. *Rapid Prototyp. J.* **2004**, *10*, 7–13. [CrossRef]
9. Noguera, R.; Lejeune, M.; Chartier, T. 3D fine scale ceramic components formed by ink-jet prototyping process. *J. Eur. Ceram. Soc.* **2005**, *25*, 2055–2059. [CrossRef]
10. Shirazi, S.F.S.; Gharehkhani, S.; Mehrali, M.; Yarmand, H.; Metselaar, H.S.C.; Adib Kadri, N.; Osman, N.A.A. A review on powder-based additive manufacturing for tissue engineering: Selective laser sintering and inkjet 3D printing. *Sci. Technol. Adv. Mater.* **2015**, *16*, 033502. [CrossRef] [PubMed]
11. Cooper, G.M.; Miller, E.D.; DeCesare, G.E.; Usas, A.; Lensie, E.L.; Bykowski, M.R.; Huard, J.; Weiss, L.E.; Losee, J.E.; Campbell, P.G. Inkjet-based biopatterning of bone morphogenetic protein-2 to spatially control calvarial bone formation. *Tissue Eng. Part A* **2010**, *16*, 1749–1759. [CrossRef] [PubMed]
12. Inzana, J.A.; Olvera, D.; Fuller, S.M.; Kelly, J.P.; Graeve, O.A.; Schwarz, E.M.; Kates, S.L.; Awad, H.A. 3D printing of composite calcium phosphate and collagen scaffolds for bone regeneration. *Biomaterials* **2014**, *35*, 4026–4034. [CrossRef] [PubMed]
13. Schiele, N.R.; Corr, D.T.; Huang, Y.; Raof, N.A.; Xie, Y.; Chrisey, D.B. Laser-based direct-write techniques for cell printing. *Biofabrication* **2010**, *2*, 032001. [CrossRef] [PubMed]
14. Colina, M.; Serra, P.; Fernandez-Pradas, J.M.; Sevilla, L.; Morenza, J.L. DNA deposition through laser induced forward transfer. *Biosens. Bioelectron.* **2005**, *20*, 1638–1642. [CrossRef] [PubMed]
15. Mandrycky, C.; Wang, Z.; Kim, K.; Kim, D.H. 3D bioprinting for engineering complex tissues. *Biotechnol. Adv.* **2016**, *34*, 422–434. [CrossRef] [PubMed]
16. Mazzoli, A. Selective laser sintering in biomedical engineering. *Med Biol. Eng. Comput.* **2013**, *51*, 245–256. [CrossRef] [PubMed]
17. Brunello, G.; Sivolella, S.; Meneghello, R.; Ferroni, L.; Gardin, C.; Piattelli, A.; Zavan, B.; Bressan, E. Powder-based 3D printing for bone tissue engineering. *Biotechnol. Adv.* **2016**, *34*, 740–753. [CrossRef]

18. Roskies, M.G.; Fang, D.; Abdallah, M.-N.; Charbonneau, A.M.; Cohen, N.; Jordan, J.O.; Hier, M.P.; Mlynarek, A.; Tamimi, F.; Tran, S.D. Three-dimensionally printed polyetherketoneketone scaffolds with mesenchymal stem cells for the reconstruction of critical-sized mandibular defects. *Laryngoscope* **2017**, *127*, E392–E398. [CrossRef]
19. Roskies, M.; Jordan, J.O.; Fang, D.; Abdallah, M.N.; Hier, M.P.; Mlynarek, A.; Tamimi, F.; Tran, S.D. Improving PEEK bioactivity for craniofacial reconstruction using a 3D printed scaffold embedded with mesenchymal stem cells. *J. Biomater. Appl.* **2016**, *31*, 132–139. [CrossRef]
20. Zhang, L.; Shen, S.; Yu, H.; Shen, S.G.; Wang, X. Computer-aided design and computer-aided manufacturing hydroxyapatite/epoxide acrylate maleic compound construction for craniomaxillofacial bone defects. *J. Craniofac. Surg.* **2015**, *26*, 1477–1481. [CrossRef]
21. Rasperini, G.; Pilipchuk, S.P.; Flanagan, C.L.; Park, C.H.; Pagni, G.; Hollister, S.J.; Giannobile, W.V. 3D-printed bioresorbable scaffold for periodontal repair. *J. Dent. Res.* **2015**, *94*, 153s–157s. [CrossRef] [PubMed]
22. Nyberg, E.L.; Farris, A.L.; Hung, B.P.; Dias, M.; Garcia, J.R.; Dorafshar, A.H.; Grayson, W.L. 3D-printing technologies for craniofacial rehabilitation, reconstruction, and regeneration. *Ann. Biomed. Eng.* **2017**, *45*, 45–57. [CrossRef] [PubMed]
23. Williams, J.M.; Adewunmi, A.; Schek, R.M.; Flanagan, C.L.; Krebsbach, P.H.; Feinberg, S.E.; Hollister, S.J.; Das, S. Bone tissue engineering using polycaprolactone scaffolds fabricated via selective laser sintering. *Biomaterials* **2005**, *26*, 4817–4827. [CrossRef] [PubMed]
24. Smith, M.H.; Flanagan, C.L.; Kemppainen, J.M.; Sack, J.A.; Chung, H.; Das, S.; Hollister, S.J.; Feinberg, S.E. Computed tomography-based tissue-engineered scaffolds in craniomaxillofacial surgery. *Int. J. Med. Robot. Comput. Assist. Surg.* **2007**, *3*, 207–216. [CrossRef] [PubMed]
25. Peyre, P.; Rouchausse, Y.; Defauchy, D.; Regnier, G. Experimental and numerical analysis of the selective laser sintering (SLS) of PA12 and PEKK semi-crystalline polymers. *J. Mater. Process. Technol.* **2015**, *225*, 326–336. [CrossRef]
26. Converse, G.L.; Conrad, T.L.; Merrill, C.H.; Roeder, R.K. Hydroxyapatite whisker-reinforced polyetherketoneketone bone ingrowth scaffolds. *Acta Biomater.* **2010**, *6*, 856–863. [CrossRef] [PubMed]
27. Adamzyk, C.; Kachel, P.; Hoss, M.; Gremse, F.; Modabber, A.; Holzle, F.; Tolba, R.; Neuss, S.; Lethaus, B. Bone tissue engineering using polyetherketoneketone scaffolds combined with autologous mesenchymal stem cells in a sheep calvarial defect model. *J. Cranio-Maxillo-Fac. Surg. Off. Publ. Eur. Assoc. Cranio-Maxillo-Fac. Surg.* **2016**, *44*, 985–994. [CrossRef]
28. Lin, Y.; Umebayashi, M.; Abdallah, M.-N.; Dong, G.; Roskies, M.G.; Zhao, Y.F.; Murshed, M.; Zhang, Z.; Tran, S.D. Combination of polyetherketoneketone scaffold and human mesenchymal stem cells from temporomandibular joint synovial fluid enhances bone regeneration. *Sci. Rep.* **2019**, *9*, 472. [CrossRef]
29. Ngo, T.D.; Kashani, A.; Imbalzano, G.; Nguyen, K.T.Q.; Hui, D. Additive manufacturing (3D printing): A review of materials, methods, applications and challenges. *Compos. Part B Eng.* **2018**, *143*, 172–196. [CrossRef]
30. Wang, X.; Jiang, M.; Zhou, Z.; Gou, J.; Hui, D. 3D printing of polymer matrix composites: A review and prospective. *Compos. Part B Eng.* **2017**, *110*, 442–458. [CrossRef]
31. Li, J.; Chen, M.; Wei, X.; Hao, Y.; Wang, J. Evaluation of 3D-printed polycaprolactone scaffolds coated with freeze-dried platelet-rich plasma for bone regeneration. *Materials* **2017**, *10*, 831. [CrossRef] [PubMed]
32. Kim, K.; Lee, C.H.; Kim, B.K.; Mao, J.J. Anatomically shaped tooth and periodontal regeneration by cell homing. *J. Dent. Res.* **2010**, *89*, 842–847. [CrossRef] [PubMed]
33. Billiet, T.; Vandenhaute, M.; Schelfhout, J.; Van Vlierberghe, S.; Dubruel, P. A review of trends and limitations in hydrogel-rapid prototyping for tissue engineering. *Biomaterials* **2012**, *33*, 6020–6041. [CrossRef] [PubMed]
34. Ma, Y.; Ji, Y.; Huang, G.; Ling, K.; Zhang, X.; Xu, F. Bioprinting 3D cell-laden hydrogel microarray for screening human periodontal ligament stem cell response to extracellular matrix. *Biofabrication* **2015**, *7*, 044105. [CrossRef] [PubMed]
35. Padovani, G.C.; Feitosa, V.P.; Sauro, S.; Tay, F.R.; Durán, G.; Paula, A.J.; Durán, N. Advances in dental materials through nanotechnology: Facts, perspectives and toxicological aspects. *Trends Biotechnol.* **2015**, *33*, 621–636. [CrossRef]
36. Galler, K.M.; Hartgerink, J.D.; Cavender, A.C.; Schmalz, G.; D'Souza, R.N. A customized self-assembling peptide hydrogel for dental pulp tissue engineering. *Tissue Eng. Part A* **2011**, *18*, 176–184. [CrossRef]

37. Obregon, F.; Vaquette, C.; Ivanovski, S.; Hutmacher, D.W.; Bertassoni, L.E. Three-dimensional bioprinting for regenerative dentistry and craniofacial tissue engineering. *J. Dent. Res.* **2015**, *94*, 143s–152s. [CrossRef]
38. Stansbury, J.W.; Idacavage, M.J. 3D printing with polymers: Challenges among expanding options and opportunities. *Dent. Mater.* **2016**, *32*, 54–64. [CrossRef]
39. Tappa, K.; Jammalamadaka, U. Novel biomaterials used in medical 3D printing techniques. *J. Funct. Biomater.* **2018**, *9*, 17. [CrossRef]
40. Battistella, E.; Varoni, E.; Cochis, A.; Palazzo, B.; Rimondini, L. Degradable polymers may improve dental practice. *J. Appl. Biomater. Biomech.* **2011**, *9*, 223–231. [CrossRef]
41. Sharma, S.; Srivastava, D.; Grover, S.; Sharma, V. Biomaterials in tooth tissue engineering: a review. *J. Clin. Diagn. Res.* **2014**, *8*, 309–315. [CrossRef] [PubMed]
42. Jammalamadaka, U.; Tappa, K. Recent advances in biomaterials for 3D printing and tissue engineering. *J. Funct. Biomater.* **2018**, *9*, 22. [CrossRef] [PubMed]
43. Roohani-Esfahani, S.-I.; Newman, P.; Zreiqat, H. Design and fabrication of 3D printed scaffolds with a mechanical strength comparable to cortical bone to repair large bone defects. *Sci. Rep.* **2016**, *6*, 19468. [CrossRef] [PubMed]
44. Alluri, R.; Jakus, A.; Bougioukli, S.; Pannell, W.; Sugiyama, O.; Tang, A.; Shah, R.; Lieberman, J.R. 3D printed hyperelastic "bone" scaffolds and regional gene therapy: A novel approach to bone healing. *J. Biomed. Mater. Res. Part A* **2018**, *106*, 1104–1110. [CrossRef] [PubMed]
45. Sa, M.W.; Nguyen, B.B.; Moriarty, R.A.; Kamalitdinov, T.; Fisher, J.P.; Kim, J.Y. Fabrication and evaluation of 3D printed BCP scaffolds reinforced with ZrO_2 for bone tissue applications. *Biotechnol. Bioeng.* **2018**, *115*, 989–999. [CrossRef] [PubMed]
46. Komlev, V.S.; Popov, V.K.; Mironov, A.V.; Fedotov, A.Y.; Teterina, A.Y.; Smirnov, I.V.; Bozo, I.Y.; Rybko, V.A.; Deev, R.V. 3D printing of octacalcium phosphate bone substitutes. *Front. Bioeng. Biotechnol.* **2015**, *3*, 81. [CrossRef] [PubMed]
47. Asadi-Eydivand, M.; Solati-Hashjin, M.; Shafiei, S.S.; Mohammadi, S.; Hafezi, M.; Abu Osman, N.A. Structure, properties, and in vitro behavior of heat-treated calcium sulfate scaffolds fabricated by 3D printing. *PloS ONE* **2016**, *11*, e0151216. [CrossRef] [PubMed]
48. Zhu, W.; Xu, C.; Ma, B.P.; Zheng, Z.B.; Li, Y.L.; Ma, Q.; Wu, G.L.; Weng, X.S. Three-dimensional printed scaffolds with gelatin and platelets enhance in vitro preosteoblast growth behavior and the sustained-release effect of growth factors. *Chin. Med J.* **2016**, *129*, 2576–2581. [CrossRef] [PubMed]
49. Tsai, K.-Y.; Lin, H.-Y.; Chen, Y.-W.; Lin, C.-Y.; Hsu, T.-T.; Kao, C.-T. Laser sintered magnesium-calcium silicate/poly-ε-caprolactone scaffold for bone tissue engineering. *Materials* **2017**, *10*, 65. [CrossRef]
50. Tao, O.; Wu, D.T.; Pham, H.M.; Pandey, N.; Tran, S.D. Nanomaterials in craniofacial tissue regeneration: A review. *Appl. Sci.* **2019**, *9*, 317. [CrossRef]
51. Guvendiren, M.; Molde, J.; Soares, R.M.D.; Kohn, J. Designing biomaterials for 3D printing. *ACS Biomater. Sci. Eng.* **2016**, *2*, 1679–1693. [CrossRef] [PubMed]
52. Mironov, V.; Visconti, R.P.; Kasyanov, V.; Forgacs, G.; Drake, C.J.; Markwald, R.R. Organ printing: Tissue spheroids as building blocks. *Biomaterials* **2009**, *30*, 2164–2174. [CrossRef] [PubMed]
53. Norotte, C.; Marga, F.S.; Niklason, L.E.; Forgacs, G. Scaffold-free vascular tissue engineering using bioprinting. *Biomaterials* **2009**, *30*, 5910–5917. [CrossRef] [PubMed]
54. Munaz, A.; Vadivelu, R.K.; St. John, J.; Barton, M.; Kamble, H.; Nguyen, N.-T. Three-dimensional printing of biological matters. *J. Sci. Adv. Mater. Devices* **2016**, *1*, 1–17. [CrossRef]
55. Chung, J.H.Y.; Naficy, S.; Yue, Z.; Kapsa, R.; Quigley, A.; Moulton, S.E.; Wallace, G.G. Bio-ink properties and printability for extrusion printing living cells. *Biomater. Sci.* **2013**, *1*, 763–773. [CrossRef]
56. Tan, Y.; Richards, D.J.; Trusk, T.C.; Visconti, R.P.; Yost, M.J.; Kindy, M.S.; Drake, C.J.; Argraves, W.S.; Markwald, R.R.; Mei, Y. 3D printing facilitated scaffold-free tissue unit fabrication. *Biofabrication* **2014**, *6*, 024111. [CrossRef] [PubMed]
57. Lozano, R.; Stevens, L.; Thompson, B.C.; Gilmore, K.J.; Gorkin, R.; Stewart, E.M.; in het Panhuis, M.; Romero-Ortega, M.; Wallace, G.G. 3D printing of layered brain-like structures using peptide modified gellan gum substrates. *Biomaterials* **2015**, *67*, 264–273. [CrossRef]
58. Nakayama, K. Chapter 1—In vitro biofabrication of tissues and organs. In *Biofabrication*; Forgacs, G., Sun, W., Eds.; William Andrew Publishing: Boston, MA, USA, 2013; pp. 1–21.

59. Carter, S.D.; Costa, P.F.; Vaquette, C.; Ivanovski, S.; Hutmacher, D.W.; Malda, J. Additive biomanufacturing: An advanced approach for periodontal tissue regeneration. *Ann. Biomed. Eng.* **2017**, *45*, 12–22. [CrossRef]
60. Oberoi, G.; Nitsch, S.; Edelmayer, M.; Janjic, K.; Muller, A.S.; Agis, H. 3D printing-encompassing the facets of dentistry. *Front. Bioeng. Biotechnol.* **2018**, *6*, 172. [CrossRef]
61. Ivanovski, S.; Vaquette, C.; Gronthos, S.; Hutmacher, D.W.; Bartold, P.M. Multiphasic scaffolds for periodontal tissue engineering. *J. Dent. Res.* **2014**, *93*, 1212–1221. [CrossRef]
62. Lee, C.H.; Hajibandeh, J.; Suzuki, T.; Fan, A.; Shang, P.; Mao, J.J. Three-dimensional printed multiphase scaffolds for regeneration of periodontium complex. *Tissue Eng. Part A* **2014**, *20*, 1342–1351. [CrossRef] [PubMed]
63. Hamlet, S.M.; Vaquette, C.; Shah, A.; Hutmacher, D.W.; Ivanovski, S. 3-dimensional functionalized polycaprolactone-hyaluronic acid hydrogel constructs for bone tissue engineering. *J. Clin. Periodontol.* **2017**, *44*, 428–437. [CrossRef] [PubMed]
64. Vaquette, C.; Fan, W.; Xiao, Y.; Hamlet, S.; Hutmacher, D.W.; Ivanovski, S. A biphasic scaffold design combined with cell sheet technology for simultaneous regeneration of alveolar bone/periodontal ligament complex. *Biomaterials* **2012**, *33*, 5560–5573. [CrossRef] [PubMed]
65. Farag, A.; Hashimi, S.M.; Vaquette, C.; Bartold, P.M.; Hutmacher, D.W.; Ivanovski, S. The effect of decellularized tissue engineered constructs on periodontal regeneration. *J. Clin. Periodontol.* **2018**, *45*, 586–596. [CrossRef] [PubMed]
66. Bakirci, E.; Toprakhisar, B.; Zeybek, M.C.; Ince, G.O.; Koc, B. Cell sheet based bioink for 3D bioprinting applications. *Biofabrication* **2017**, *9*, 024105. [CrossRef]
67. Vaquette, C.; Pilipchuk, S.P.; Bartold, P.M.; Hutmacher, D.W.; Giannobile, W.V.; Ivanovski, S. Tissue engineered constructs for periodontal regeneration: Current status and future perspectives. *Adv. Healthc. Mater.* **2018**, *7*, 1800457. [CrossRef]
68. Li, X.; Ma, C.; Xie, X.; Sun, H.; Liu, X. Pulp regeneration in a full-length human tooth root using a hierarchical nanofibrous microsphere system. *Acta Biomater.* **2016**, *35*, 57–67. [CrossRef]
69. Murray, P.E.; Garcia-Godoy, F.; Hargreaves, K.M. Regenerative endodontics: A review of current status and a call for action. *J. Endod.* **2007**, *33*, 377–390. [CrossRef]
70. Ma, Y.; Xie, L.; Yang, B.; Tian, W. Three-dimensional printing biotechnology for the regeneration of the tooth and tooth-supporting tissues. *Biotechnol. Bioeng.* **2019**, *116*, 452–468. [CrossRef]
71. Athirasala, A.; Tahayeri, A.; Thrivikraman, G.; Franca, C.M.; Monteiro, N.; Tran, V.; Ferracane, J.; Bertassoni, L.E. A dentin-derived hydrogel bioink for 3D bioprinting of cell laden scaffolds for regenerative dentistry. *Biofabrication* **2018**, *10*, 024101. [CrossRef]
72. Feng, K.-C.; Pinkas-Sarafova, A.; Ricotta, V.; Cuiffo, M.; Zhang, L.; Guo, Y.; Chang, C.-C.; Halada, G.P.; Simon, M.; Rafailovich, M. The influence of roughness on stem cell differentiation using 3D printed polylactic acid scaffolds. *Soft Matter.* **2018**, *14*, 9838–9846. [CrossRef] [PubMed]
73. Hu, Y.; Wu, Y.; Gou, Z.; Tao, J.; Zhang, J.; Liu, Q.; Kang, T.; Jiang, S.; Huang, S.; He, J.; et al. 3D-engineering of cellularized conduits for peripheral nerve regeneration. *Sci. Rep.* **2016**, *6*, 32184. [CrossRef] [PubMed]
74. Lobo, S.E.; Glickman, R.; da Silva, W.N.; Arinzeh, T.L.; Kerkis, I. Response of stem cells from different origins to biphasic calcium phosphate bioceramics. *Cell Tissue Res.* **2015**, *361*, 477–495. [CrossRef] [PubMed]
75. Saijo, H.; Igawa, K.; Kanno, Y.; Mori, Y.; Kondo, K.; Shimizu, K.; Suzuki, S.; Chikazu, D.; Iino, M.; Anzai, M.; et al. Maxillofacial reconstruction using custom-made artificial bones fabricated by inkjet printing technology. *J. Artif. Organs Off. J. Jpn. Soc. Artif. Organs* **2009**, *12*, 200–205. [CrossRef] [PubMed]
76. Shao, H.; Sun, M.; Zhang, F.; Liu, A.; He, Y.; Fu, J.; Yang, X.; Wang, H.; Gou, Z. Custom repair of mandibular bone defects with 3D printed bioceramic scaffolds. *J. Dent. Res.* **2018**, *97*, 68–76. [CrossRef] [PubMed]
77. Kargozar, S.; Baino, F.; Hamzehlou, S.; Hill, R.G.; Mozafari, M. Bioactive glasses: Sprouting angiogenesis in tissue engineering. *Trends Biotechnol.* **2018**, *36*, 430–444. [CrossRef] [PubMed]
78. Park, S.H.; Park, D.S.; Shin, J.W.; Kang, Y.G.; Kim, H.K.; Yoon, T.R.; Shin, J.W. Scaffolds for bone tissue engineering fabricated from two different materials by the rapid prototyping technique: PCL versus PLGA. *J. Mater. Science. Mater. Med.* **2012**, *23*, 2671–2678. [CrossRef]
79. Shim, J.H.; Yoon, M.C.; Jeong, C.M.; Jang, J.; Jeong, S.I.; Cho, D.W.; Huh, J.B. Efficacy of rhBMP-2 loaded PCL/PLGA/beta-TCP guided bone regeneration membrane fabricated by 3D printing technology for reconstruction of calvaria defects in rabbit. *Biomed. Mater.* **2014**, *9*, 065006. [CrossRef]

80. Hung, B.P.; Naved, B.A.; Nyberg, E.L.; Dias, M.; Holmes, C.A.; Elisseeff, J.H.; Dorafshar, A.H.; Grayson, W.L. Three-dimensional printing of bone extracellular matrix for craniofacial regeneration. *ACS Biomater. Sci. Eng.* **2016**, *2*, 1806–1816. [CrossRef]
81. Fedore, C.W.; Tse, L.Y.L.; Nam, H.K.; Barton, K.L.; Hatch, N.E. Analysis of polycaprolactone scaffolds fabricated via precision extrusion deposition for control of craniofacial tissue mineralization. *Orthod. Craniofacial Res.* **2017**, *20* (Suppl. 1), 12–17. [CrossRef]
82. Rodríguez-Méndez, I.; Fernández-Gutiérrez, M.; Rodríguez-Navarrete, A.; Rosales-Ibáñez, R.; Benito-Garzón, L.; Vázquez-Lasa, B.; San Román, J. Bioactive Sr(II)/Chitosan/Poly(ε-caprolactone) scaffolds for craniofacial tissue regeneration. In vitro and in vivo behavior. *Polymers* **2018**, *10*, 279. [CrossRef] [PubMed]
83. Daly, A.C.; Cunniffe, G.M.; Sathy, B.N.; Jeon, O.; Alsberg, E.; Kelly, D.J. 3D bioprinting of developmentally inspired templates for whole bone organ engineering. *Adv. Healthc. Mater* **2016**, *5*, 2353–2362. [CrossRef] [PubMed]
84. Rhee, S.; Puetzer, J.L.; Mason, B.N.; Reinhart-King, C.A.; Bonassar, L.J. 3D bioprinting of spatially heterogeneous collagen constructs for cartilage tissue engineering. *ACS Biomater. Sci. Eng.* **2016**, *2*, 1800–1805. [CrossRef]
85. Schuurman, W.; Levett, P.A.; Pot, M.W.; van Weeren, P.R.; Dhert, W.J.; Hutmacher, D.W.; Melchels, F.P.; Klein, T.J.; Malda, J. Gelatin-methacrylamide hydrogels as potential biomaterials for fabrication of tissue-engineered cartilage constructs. *Macromol. Biosci.* **2013**, *13*, 551–561. [CrossRef] [PubMed]
86. Zopf, D.A.; Mitsak, A.G.; Flanagan, C.L.; Wheeler, M.; Green, G.E.; Hollister, S.J. Computer aided-designed, 3-dimensionally printed porous tissue bioscaffolds for craniofacial soft tissue reconstruction. *Otolaryngol. Head Neck Surg.* **2015**, *152*, 57–62. [CrossRef] [PubMed]
87. Morrison, R.J.; Nasser, H.B.; Kashlan, K.N.; Zopf, D.A.; Milner, D.J.; Flanangan, C.L.; Wheeler, M.B.; Green, G.E.; Hollister, S.J. Co-culture of adipose-derived stem cells and chondrocytes on three-dimensionally printed bioscaffolds for craniofacial cartilage engineering. *Laryngoscope* **2018**, *128*, E251–E257. [CrossRef] [PubMed]
88. Lee, J.S.; Hong, J.M.; Jung, J.W.; Shim, J.H.; Oh, J.H.; Cho, D.W. 3D printing of composite tissue with complex shape applied to ear regeneration. *Biofabrication* **2014**, *6*, 024103. [CrossRef] [PubMed]
89. Park, S.H.; Yun, B.G.; Won, J.Y.; Yun, W.S.; Shim, J.H.; Lim, M.H.; Kim, D.H.; Baek, S.A.; Alahmari, Y.D.; Jeun, J.H.; et al. New application of three-dimensional printing biomaterial in nasal reconstruction. *Laryngoscope* **2017**, *127*, 1036–1043. [CrossRef]
90. Kang, H.-W.; Lee, S.J.; Ko, I.K.; Kengla, C.; Yoo, J.J.; Atala, A. A 3D bioprinting system to produce human-scale tissue constructs with structural integrity. *Nat. Biotechnol.* **2016**, *34*, 312. [CrossRef]
91. Seyednejad, H.; Gawlitta, D.; Kuiper, R.V.; de Bruin, A.; van Nostrum, C.F.; Vermonden, T.; Dhert, W.J.; Hennink, W.E. In vivo biocompatibility and biodegradation of 3D-printed porous scaffolds based on a hydroxyl-functionalized poly(epsilon-caprolactone). *Biomaterials* **2012**, *33*, 4309–4318. [CrossRef]
92. Tarafder, S.; Koch, A.; Jun, Y.; Chou, C.; Awadallah, M.R.; Lee, C.H. Micro-precise spatiotemporal delivery system embedded in 3D printing for complex tissue regeneration. *Biofabrication* **2016**, *8*, 025003. [CrossRef] [PubMed]
93. Legemate, K.; Tarafder, S.; Jun, Y.; Lee, C.H. Engineering human TMJ discs with protein-releasing 3D-printed scaffolds. *J. Dent. Res.* **2016**, *95*, 800–807. [CrossRef] [PubMed]
94. Visser, J.; Melchels, F.P.; Jeon, J.E.; van Bussel, E.M.; Kimpton, L.S.; Byrne, H.M.; Dhert, W.J.; Dalton, P.D.; Hutmacher, D.W.; Malda, J. Reinforcement of hydrogels using three-dimensionally printed microfibres. *Nat. Commun.* **2015**, *6*, 6933. [CrossRef] [PubMed]

Review

In Vivo Tracking of Tissue Engineered Constructs

Carmen J. Gil [1], Martin L. Tomov [1], Andrea S. Theus [1], Alexander Cetnar [1], Morteza Mahmoudi [2,3] and Vahid Serpooshan [1,4,5,*]

1 Wallace H. Coulter Department of Biomedical Engineering, Emory University School of Medicine and Georgia Institute of Technology, Atlanta, GA 30322, USA
2 Precision Health Program, Michigan State University, East Lansing, MI 48824, USA
3 Department of Radiology, Michigan State University, East Lansing, MI 48824, USA
4 Department of Pediatrics, Emory University School of Medicine, Atlanta, GA 30309, USA
5 Children's Healthcare of Atlanta, Atlanta, GA 30322, USA
* Correspondence: vahid.serpooshan@bme.gatech.edu; Tel.: +404-712-9717

Received: 2 July 2019; Accepted: 13 July 2019; Published: 16 July 2019

Abstract: To date, the fields of biomaterials science and tissue engineering have shown great promise in creating bioartificial tissues and organs for use in a variety of regenerative medicine applications. With the emergence of new technologies such as additive biomanufacturing and 3D bioprinting, increasingly complex tissue constructs are being fabricated to fulfill the desired patient-specific requirements. Fundamental to the further advancement of this field is the design and development of imaging modalities that can enable visualization of the bioengineered constructs following implantation, at adequate spatial and temporal resolution and high penetration depths. These in vivo tracking techniques should introduce minimum toxicity, disruption, and destruction to treated tissues, while generating clinically relevant signal-to-noise ratios. This article reviews the imaging techniques that are currently being adopted in both research and clinical studies to track tissue engineering scaffolds in vivo, with special attention to 3D bioprinted tissue constructs.

Keywords: in vivo imaging; tissue engineering; 3D bioprinting; additive manufacturing; scaffold tracking; magnetic resonant imaging (MRI); computed tomography (CT); ultrasound; fluorescence spectroscopy; bioluminescence; optical coherence tomography; photoacoustic imaging; magnetic-particle imaging; multimodal imaging

---

1. Introduction

A significant portion of recent advancements in the field of tissue engineering (TE) has focused on design, developing, and characterization of new biomaterials that can be used as tissue mimics to model a variety of diseases in vitro, or as implants to repair or regenerate damaged tissues in vivo [1–3]. Further, the advent of new automated additive manufacturing techniques, such as 3D printing and bioprinting, together with computer-aided design (CAD) modeling, have allowed for higher throughput biofabrication of 3D scaffolding systems with increasing structural and functional complexities to be used in patient-specific TE and precision medicine applications [4–8]. Thus, it is vital to design and utilize effective imaging and tracking methods to closely monitor the scaffolds following implantation in the patient's body [9,10]. These techniques should enable noninvasive, real-time examination of properties including the graft stability and position, biomaterial-tissue interactions (e.g., biocompatibility, degradation, and integration with host tissue), blood perfusion (angiogenesis), and function (e.g., contractile function of a cardiac patch). To achieve this goal, imaging techniques with minimal invasiveness as well high penetration depth and high resolution are required to provide a clear contrast between the embedded biological materials and the surrounding tissue structure, thus generating a complete picture covering morphological, physiological, and molecular processes [11].

New advancements in medical imaging have been made to address different challenges in the TE field, ranging from the design to production processes of tissues and organs, as well as clinical implantation and implementation (Figure 1) [10,12]. These imaging systems often follow a common process of exciting the targeted samples with an energy source like electromagnetic radiation, light, or sound, or a combination of those sources, to generate a response in the form of emitted, transmitted or reflected signal which can then be captured through different detector designs for analysis [10]. Furthermore, specific techniques also require contrast agents or sample labeling methods to enhance the signal-to-noise ratio [13]. As a result, different methods would have distinct advantages and disadvantages with respect to the penetration depth, image temporal and spatial resolution, as well as the effect of the exciting source and contrast/labeling agent on the biological target(s) (Table 1). Thus, it is important that the techniques are selected carefully and tailored to fulfill the specific application requirements.

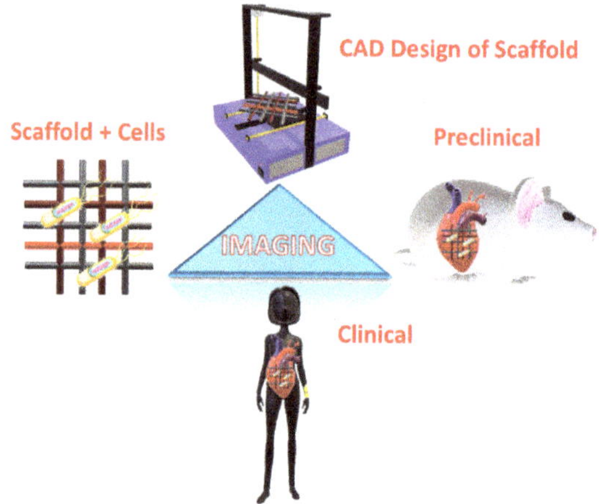

Figure 1. Schematic illustration for the role of imaging in tissue engineering (TE) applications at different levels: scaffold design using computer-aided design (CAD), cellular scaffolds in in vitro applications, preclinical application through implantation in animal models, and clinical application in humans.

In this review, we detail the different medical imaging techniques used in TE applications. These methods include: computed tomography (CT), magnetic resonance imaging (MRI), magnetic particle imaging (MPI), ultrasound, photoacoustic imaging, different optical methods (fluorescence spectroscopy, bioluminescence, and optical coherence tomography), and multimodal imaging (Table 1). These methods have been widely applied in the different areas of TE for investigating the morphological structures of the scaffold structures as well as studying the viability of the different cellular constructs [11]. The Main focus will be on delineating how the general principle of operation and recent advancements in the imaging methods can aid the researchers in the field to select the most effective imaging methods for their in vivo studies.

Table 1. List of non-invasive imaging methods, their resolution and depth, costs, external material usage, information type as well as applications in TE ranging from imaging scaffolds (with or without cells), and preclinical and clinical applications.

Method	Spatial Resolution	Imaging Depth	Information Type	Cost	External Material	Applications
CT	5 µm	No limit	3D	Medium	Yes	Scaffold + cells + Pre/Clinical
MRI	5–200 µm	No limit	3D	High	No	Scaffold + cells + Pre/Clinical
MPI	1 µm	No limit	2D/3D	Medium	Yes	Scaffold + cells + Pre/Clinical
Ultrasound	20–100 µm	10 mm	3D	Medium	No	Scaffold + cells + Pre/Clinical
Fluorescence	0.2–1 µm	0.3–1 mm	2D	Low	Yes	Scaffold + cells
Bioluminescence	2–3 mm	10 mm	2D	Low	Yes	Scaffold + cells
Optical coherence tomography (OCT)	1–15 µm	1–3 mm	2D	Low	No	Scaffold + cells + Pre/Clinical
Photoacoustic	50–150 µm	20 mm	2D/3D	Medium	No	Scaffold + cells + Pre/Clinical

2. Scaffold Tracking Techniques

2.1. Magnetic Resonance Imaging (MRI)

Magnetic resonance imaging (MRI) is a commonly used imaging method that has broad applications in the clinical and basic research fields. Briefly, MRI is a non-invasive imaging technique that allows for pertinent information to be gathered over the entire patient body in a highly detailed manner [14]. Importantly, MRI is usually not associated with harmful radiations, allowing for repetitive scans and longitudinal studies to be performed with minimal harmful effects, which further makes this imaging method an attractive way to diagnose clinical pathologies [15]. It uses magnetic fields and radio waves often combined with contrast agents to generate highly detailed images of tissues within the body and is widely used for applications in the clinics. Recently, there has been an increasing interest for the use of MRI in various TE applications, because of the capability of the technique in producing high-resolution 3D structural scans with minimal damage to the tissue mimic [16–18]. Such 3D reconstructions can be used, via 3D bioprinting methods, to create patient and disease-specific tissue constructs for regenerative therapies (Figure 2).

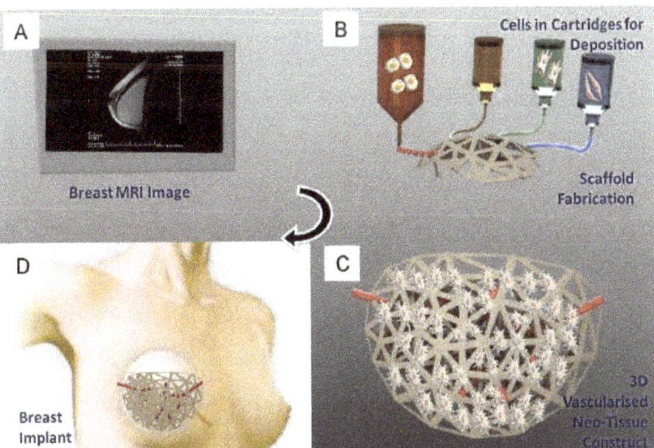

Figure 2. Application of patient's MRI data to generate a bioprinted scaffold for organ regeneration, disease treatment, or drug delivery. MR images (**A**) of the target organ/tissue will be acquired and processed to create a 3D STL file. The model will be 3D bioprinted using various inks and scaffolds (**B**), cultured in vitro to establish the new tissue structure and vasculature (**C**), followed by implantation in vivo to repair/regenerate target tissue/organ (**D**). Reproduced with permission from Ref. [19].

Recent advances in contrast agent design and development have made it possible to detect and track cell populations within the complex tissue-engineered constructs both in vitro and post implantation in vivo. These in vivo MRI cell-tracking processes can be performed using a variety of contrast agents, such as gadolinium, fluorine, or manganese, as well as superparamagnetic nanoparticles [20,21]. These materials are the preferred MRI contrast agents for TE applications as they are usually less cytotoxic and offer more reliable cellular uptake for imaging. In addition, MRI used in conjunction with different functionalized nano-contrast agents and other imaging techniques has also been used to study different drug release kinetics in the field of tissue engineering [22–24]. To date, MRI imaging is increasingly used in conjunction with the next-generation additive manufacturing technologies, such as bioprinting, to track cells in TE scaffolds. MRI has been successfully used as both a diagnostic and tracking tool, which readily allows for translation of in vitro imaging processes in the basic research stage to the clinical settings [19,25].

For instance, superparamagnetic nano-sized iron oxide particles, coated with polyethylene glycol (PEG), have been used to label both rat and human T-cells in vivo, with over 90% efficiency and without any measurable effects on T-cell properties [26]. Iron oxide nanoparticles were used in another study as an MRI contrast agent, to label and track collagen-based cardiac patches following implantation onto the epicardial surface of the mouse heart (Figure 3) [17,18,27,28]. T2*-weighted MR images demonstrated the robust capability of this technique to noninvasively visualize the engineered patch device.

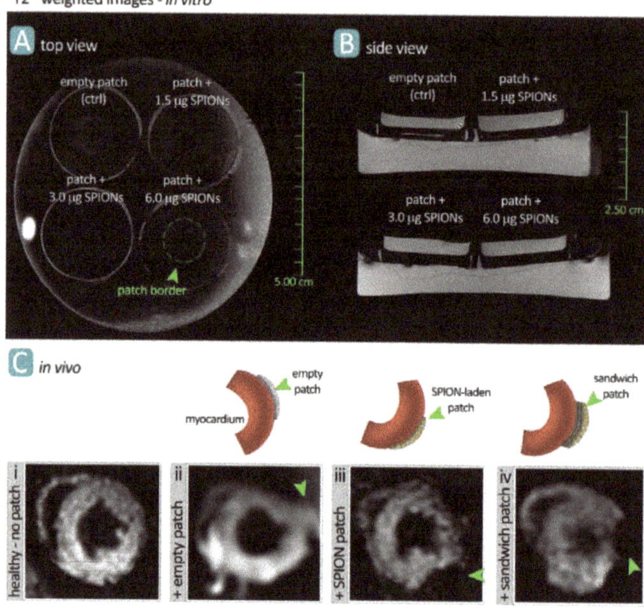

Figure 3. MR imaging of bioengineered collagen constructs used as cardiac patch to repair ischemic heart tissue. Patches were loaded with 1.5, 3.0, and 6.0 µg/mL of iron oxide nanoparticles and imaged via MRI both in vitro (**A**,**B**) and in vivo (**C**), in a mouse model. Manganese-enhanced MRI visualized the patch grafted onto healthy myocardial tissue in different groups including no treatment (control) (**i**), empty patch (**ii**), nanoparticle-loaded patch (**iii**), and loaded-empty-loaded sandwich patch (**iv**). Reproduced with permission from Ref. [17].

Stem cell-derived cellular cultures used in cartilage [29], adipose [30] and heart TE [31] have also successfully used MRI in the development, characterization, and clinical translation of the scaffolding constructs. Furthermore, MRI imaging with superparamagnetic iron oxide nanoparticles as exogenous labeling agents has been used to study different stem cell dynamics for both preclinical and clinical applications [32]. As a result, with the rapid growth of various stem cell therapies, clinical MRI will be more extensively used as a robust, noninvasive bedside tool for guided administration, delivery, and tracking of transplanted cells [33]. There is also a small but significant body of work that applied MRI imaging to nanoparticle vaccine efficacy, focusing on immune system priming and cellular activation in cancer vaccine development [34].

2.2. *Magnetic Particle Imaging (MPI)*

Apart from MRI, superparamagnetic iron oxide nanoparticles (SPIONs) have also been used in an emergent imaging technique such as magnetic particle imaging (MPI) through their strong magnetization (Figure 4) [35–37]. MPI, in conjunction with MRI as a paired technique, is gaining traction in clinical diagnostics due to its significant benefits over other more established techniques [36–38]. Specifically, MPI is a relatively fast imaging method, generates zero tissue background signal, and there is no attenuation of the signal correlated to organ depth, allowing for unimpeded and quantitative high-resolution imaging at any depth and location [38–40]. Recent work has advanced the technique in its potential for robust and sensitive cardiovascular imaging of healthy and diseased conditions, such as stenosis and myocardial infarcts [39,41,42] and cell tracking [43–47], which is of great interest in the field of TE. In particular, MPI can be useful in imaging bioprinted organ constructs, where 3D spatial arrangement and resolution are fundamental limitations. Further progress in MPI, as a diagnostic and basic research tool, requires advancements in imaging physics, nanoparticle synthesis, and characterization, as well as ongoing proof-of-principle imaging of small animals, TE constructs, and more broadly in human patients [48].

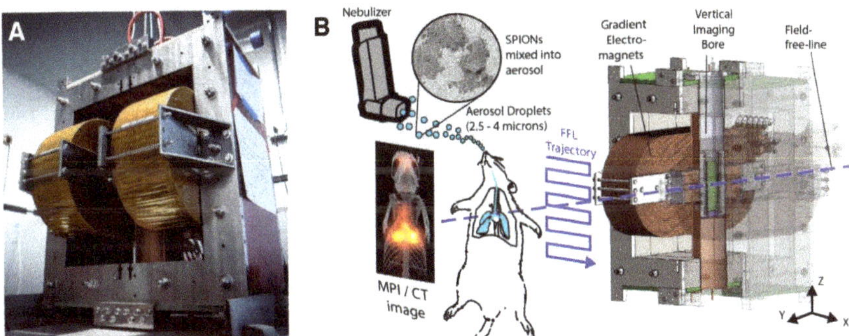

Figure 4. Hardware setup used in magnetic particle imaging (MPI) scans using SPIONs. (**A**) The Berkeley field-free-line MPI preclinical scanner. (**B**) To form a projection image, the magnetic field (FFL) rasters across a trajectory as shown, imaging the in vivo distribution of SPIONs in a rat. Multiple such projections can reconstruct a 3D MPI image similar to CT. Reproduced with permission from Ref. [48].

MPI works are based on the direct imaging of the concentration and location of the SPION tracers, using varying magnetic fields, which have high sensitivity and are capable of significant background to signal (contrast) resolution. The field has already advanced several particles (e.g., Ferumoxytol) through the FDA for chronic kidney disease induced anemia treatment [49]. Additionally, SPIONs have been shown to successfully work in patient imaging (Resovist) [50,51], and have been used to localize sentinel lymph nodes for breast cancer detection (Sienna) [52]. They have been also used for evaluation of hyperthermia-induced solid tumor removal (NanoTherm) [53]. Importantly, MPI scans are relatively safe and radiation-free, which combined with the high contrast and sensitivity imaging capabilities, offers critical advantages in cardiovascular imaging and cell tracking. MPI has been used both as a diagnostic tool and as an imaging platform in bioprinting-based TE, to build tissue/organ mimics with high fidelity to their living analogs. It is capable of detecting low numbers of cells reliably [54], which opens up the technology to be used in cell-based clinical applications, such as cancer therapies and tracking stem cell-derived TE constructs [43,46,52]. Unlike traditional radiotracing contrasts, SPIONs have a half-life that is essentially unlimited, which allows researchers to track cellular localization over long time intervals (up to several months in animal models) [55].

2.3. Angiography

An angiogram or arteriogram is a diagnostic procedure that uses specific dyes to outline the arteries in a patient (Figure 5). Arteries are invisible to the clinical imaging tools under normal conditions, and thus, their visualization requires utilizing some type of contrast agents. There are three main forms of angiograms, each relating to the imaging platform that is used to generate the clinical images.

Digital subtraction angiography (DSA) is the more common method of getting arterial images in the clinic. The artery to be imaged is numbed with a local anesthetic and then a contrast agent is injected, which outlines the vascular network downstream. Following contrast introduction, X-ray is used to acquire the vasculature images [56–59]. DSA normally takes around 20 min to perform. Computer tomography (CT) angiography is another method to acquire high-resolution 3D images of a patient's vasculature. Similar to DSA, CT angiography also requires a contrast agent introduction, but unlike DSA, the injection site is the vein in the arm, usually a drip, which allows for the entire arterial network to be imaged, if required. Image acquisition is very fast as it only takes a few seconds to generate them. The third common angiography method uses MR to generate the needed 3D vasculature images [58,60–64]. Gadolinium is the most common contrast agent that is used with MR angiography (Figure 3). Post introduction, any artery in the body can be imaged. As with CT angiography, MR approach is a fast procedure, usually performed on the same day.

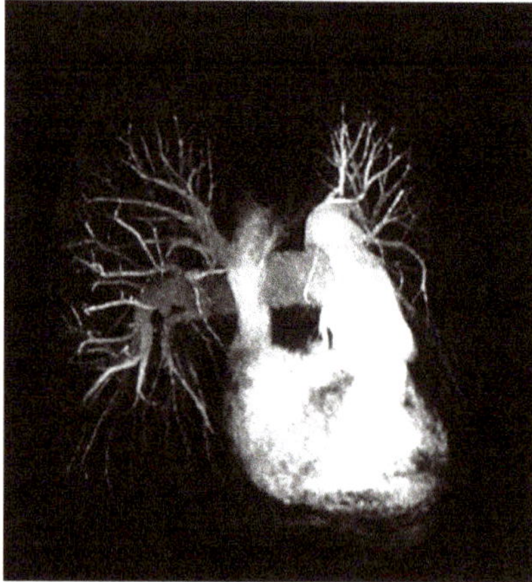

Figure 5. Example angiography (MR) outlining via contrasting the heart chambers and attendant vasculature. Reproduced with permission from Ref. [64].

Since angiography procedures involve the injection of a contrast agent to generate images, there may be some risks, such as allergies to the contrast, bleeding at the puncture site, or false aneurism [65]. A rarer but serious complication can happen if there is already some kidney damage present, where contrast injection can further deteriorate kidney function. Each of these complications can be successfully mitigated via appropriate pre-procedure preparations, or with simple surgical post-procedure manipulation in the case of the false aneurysm.

Angiography is almost exclusively a clinical imaging technique, so its application to in vitro TE has been limited so far, done predominantly in excised tissue slices from mice and pig. Nonetheless, as the additive manufacturing, specifically bioprinting technologies, enter the tissue bioengineering field, complex in vitro tissue models that incorporate vascularization will require advanced visualization and tracking methods for both modeling applications in the lab and for translational applications, such as cardiac patch implants or vessel grafts post-stenosis. Having a complete picture of all sources of flow into and out of bioengineered tissue mimics would be critical to recapitulating their functionality.

2.4. Computed Tomography (CT)

CT has been widely used as a biomedical imaging technique over the last decades due to its high spatial and temporal resolution. CT imaging generates a 3D reconstruction of the targeted sample by collecting the transmitted X-ray at different angles using a multi-array detector (Figure 6) [12]. Since the CT contrast is sensitive to the materials that attenuate the X-ray transmission, this technique has been widely used to image tissue structures which have high mineral concentrations, like bone and the surrounding tissue. Consequently, CT has been extensively used in different bone TE applications [66–69]. The development of more sensitive techniques like micro-CT has allowed for the study of the morphology and 3D structure of different scaffold geometries in the sub-micron scale as well as the tracking of different cells incorporated into the scaffold structures [70–76]. These unique advantages have also allowed CT to be used in conjunction with new additive manufacturing techniques, such as 3D (bio)printing, for different implant-manufacturing purposes [66].

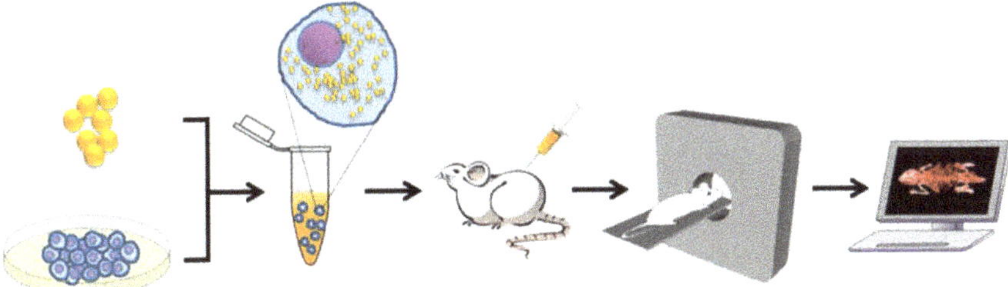

Figure 6. Cellular tracking using gold nanoparticles as a contrast agent and imaged with CT. Reproduced with permission from Ref. [77].

A drawback of the CT technique is that it is less sensitive to visualize the contrast between different soft tissue structures. However, the sensitivity can be improved by utilizing different contrasting agents. Currently, different biomaterials including gold [78–83], heavy elements [84], cationic agents [85], polymers [74], and nanoparticles [84,86] are being used as contrast agents in different CT imaging applications ranging from animal models to clinical studies. For instance, different animal model studies have employed CT with radio-transparent contrast agents like polymer [87] and alkaline-based agents [88] as in vivo imaging techniques to quantify different soft tissue structures like hepatic vascular and parenchymal regeneration as well as vascular network at a capillary level. CT with contrast angiography has also been used to study the stability of human cell-derived engineered heart valve after implantation in sheep [89,90]. Clinically, CT with an iodine-based agent has been used to study myocardial fibrosis in patients with hypertrophic cardiomyopathy [90]. These materials can be implanted in the soft tissue scaffold, thus allowing for new in vivo tracking applications. Furthermore, these nanobiomaterials can also be used as therapeutics by including functionalized medicine through surface modification [77,91]. In addition to the development of new biocompatible contrasting materials, more sensitive detectors such as photon-counting detector technology have also been developed to enhance the visualization of different soft tissues within the CT techniques [92]. As a result, contrast agents can be used both as diagnosis and therapeutics. Thus, CT imaging technologies can be used to monitor the efficacy of drug implants in vivo due to its high spatial resolution and penetration in comparison with other imaging techniques.

More recently, CT imaging has been utilized as a nondestructive tool for longitudinal and volumetric measurement of scaffold degradation both in vitro and in vivo [81]. For this purpose, gold nanoparticles, used as contract agents, were covalently conjugated to collagen polymer during scaffold fabrication, resulting in the generation of CT-visible collagen constructs. The X-ray attenuation of the conjugated scaffolds was used to measure hydrogel degradation over the time in culture.

2.5. Ultrasound

The rapid development of implantable TE platforms has created a major necessity for non-invasive, non-ionizing, and non-destructive techniques for the in vivo tracking and imaging of implantable tissues. Ultrasound imaging technologies and their associated multi-modality approaches (e.g., ultrasound-photoacoustic imaging [93]) have been investigated due to their specific advantages for TE applications (Figure 6). Importantly, ultrasound techniques can enable in-situ quantitative measurements of various properties of engineered tissues, including extracellular matrix (ECM) formation, degradation, mechanical strength, cell infiltration, vascularization, and blood perfusion and oxygenation [93,94]. This is while most in vitro or ex-vivo scaffold characterization modalities, such as electron or optical microscopy, and X-ray tomography have limitations for in vivo tracking of scaffolds, due to their invasiveness, limited penetration depth (few hundred micrometers), or poor contrast. Thus, ultrasound could be an ideal tool for diverse preclinical and clinical applications.

Ultrasound utilizes sound waves at frequencies over 20 kHz (Figure 7). In a clinical setting, frequencies ranging from 1–15 MHz are used to generate images of features in biological tissues [95]. For instance, PEG hydrogels have been characterized using B-mode ultrasound to visualize varying amounts of ECM proteins and cell composition in real time over 18 days [96]. This technique utilizes a 12 MHz imaging frequency which limits the penetration, making it difficult for probing scaffolds in vivo. However, this method is suitable for validation experiments in preclinical applications. This validation method was used in a similar study, where collagen deposition was calculated in scaffolds with myofibroblasts to quantify protein concentration [94,97–99].

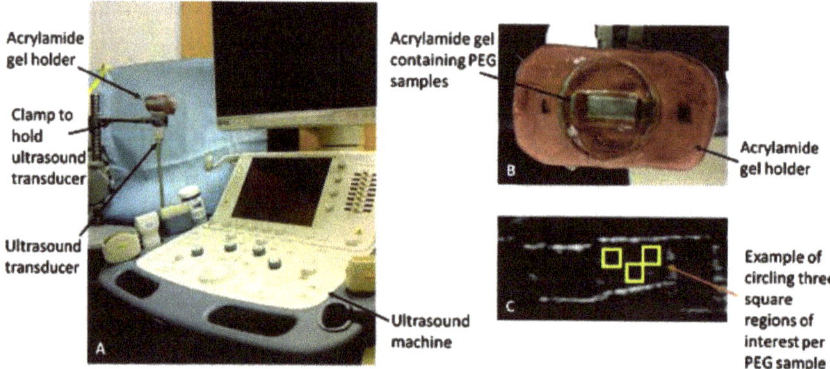

Figure 7. (**A**,**B**) Experimental setup of ultrasound used to image different areas of a scaffold made of PEG hydrogel. (**C**) Example of output image from the system. Reproduced with permission from Ref. [96].

Along with quantifying biological components concentration, mechanical properties of TE constructs can be evaluated using ultrasonic modalities coupled with computational methods. Previous studies have shown that utilizing ultrasound elastography can yield measurements of elasticity and stiffness of soft tissues to determine pathological conditions such as inflammation and tumors.

For instance, Walker et al., reported a strong correlation between the compressive moduli calculated from ultrasound and mechanical testing of TE cartilage [100]. Ultrasonic modalities offer robust information due to their unique interactions with biological tissues. As effective signal processing and computational methods improve, more methods are increasingly being applied to generate information-rich data sets of TE tissues from ultrasound acquisitions.

While ultrasound (sonography) methods may offer much lower resolution than MRI and CT for imaging of bioprinted constructs, they can monitor the condition of bioprinted tissues in vivo in real time readily, at much lower costs [101]. Ultrasound has recently found other novel applications in 3D bioprinting technologies, helping to tackle some of the main challenges related to this additive biomanufacturing techniques. Acoustic radiation in an ultrasound standing wave field (USWF) forces the cells and accumulates them at the pressure nodes, at low pressure areas [101]. This technique (so called ultrasound-assisted biofabrication) results in the formation of cell spheroids within minutes at relatively narrow/homogeneous size distributions. For instance, USWF was utilized to generate endothelial cell spheroids which showed enhanced neovessel formation [102]. Further, low-intensity ultrasound is reported to enhance stem cells proliferation and differentiation [103].

2.6. Bioluminescence

Bioluminescence is a natural light-emitting process, produced by various organisms, which can provide a measure of localization and viability of cells and tissues [104–106]. Derived from bioluminescent species, the light-emitting oxidation reaction of luciferase (enzyme) catalysis of luciferin (substrate) can be introduced to cells. Once a substrate is introduced, transfection of luciferase-coding genes into cells primes them for light production. The light intensity can then be detected and quantitatively evaluated—known as bioluminescence imaging (BLI) [104,107]. Resolutions of BLI can accurately trace down to the molecular or cellular scale across entire organisms during in vivo tracking [104,105]. Unlike potentially harmful contrast agents, this illumination process is non-invasive, biologically compatible, and can be longitudinally monitored across cell lineages. It also avoids the high cost, low throughput, and low sensitivity of standard instrumentation, such as MRI or CT. However, current challenges include deep tissue visualization, spatial resolution, homogeneous substrate distribution, and accurate interpretation of detected signals [104,108].

Cell tracking via BLI has been used for stem cells, immune cells, and bacteria for varied tissue types [104]. For instance, using human mesoangioblasts, seeded onto decellularized esophageal scaffolds, Crowley et al. quantified cell viability, proliferation, and migration after implantation in murine models over the course of 7 days [109]. Iwano et al. visualized tumorigenic cells in deep mouse lung vasculatures and also demonstrated successful tracking of hippocampal neuronal activity, while previous luciferins were too large to penetrate [110]. Notably, these longitudinal studies have been carried out for up to 16 months in marmosets [111]. To study the immune response, Conradi et al. monitored a fibrin scaffold seeded with neonatal rat heart cells when implanted in allogeneic, syngeneic, and immunodeficient rat recipients [108]. Allogenic grafts only survived for 14 ± 1 days, while the syngeneic and immunodeficient recipients lasted over 100 days, indicating the importance of autologous cell sourcing. Ex-vivo validation and improvement of BLI is continually being performed to advance in vivo techniques as well, such as stem cell seeding on intervertebral discs in culture [112]. Therefore, BLI is known as a longitudinal, non-invasive method of tracking implanted tissues and their progeny in vivo.

To improve the tissue penetration in vivo, synthetic enzyme and substrate analogues have emerged that emit longer, near-infrared (NIR) wavelengths (λ = 650–900 nm) that are unhindered by hemoglobin and melanin absorption ranges ($\lambda \leq$ 600 nm) [110]. Used luciferases originate from fireflies, sea pansies, and photobacteria, but have relatively similar maximum emission spectrums ($\lambda \leq$ 600 nm) [104]. Luciferin analogues have also been explored. For instance, AkaLumine-HCl has achieved NIR wavelengths (λ = 677 nm) and demonstrated improved spatial sensitivity in deep lung metastases down to the single cell level, as compared to luciferin-D or CycLuc1 [110]. A mutagenic derived luciferase, specific to AkaLumine-HCl, called Akaluc, was also engineered to boost catalysis efficiency by sevenfold. AkaLumine-HCl is permeable through the blood-brain barrier and evenly distributes at low concentrations [111]. Overall, as techniques evolve, the BLI utility in localizing tissue-engineered constructs longitudinally will continue to distinguish this method from other in vivo imaging modalities.

In a recent approach, functionalized bioinks were developed by incorporating luminescent optical sensor nanoparticles into the hydrogel ink solution [113]. Excitation of these nanoparticles with blue light results in the emission of red luminescent light by the particles which is in proportion to the local oxygen concentration. Higher oxygen contents will generate less red luminescence. Therefore, this innovative, noninvasive approach enables imaging and analysis of the heatmap of red luminescence and oxygen concentration within bioprinted tissue constructs (Figure 8) [113]. In another study, the proliferation of bioprinted mesenchymal stromal cells, associated with collagen and nano-hydroxyapatite, was assessed by quantification of the luciferase signal of luciferase positive cells in a mouse model for up to 42 days [114].

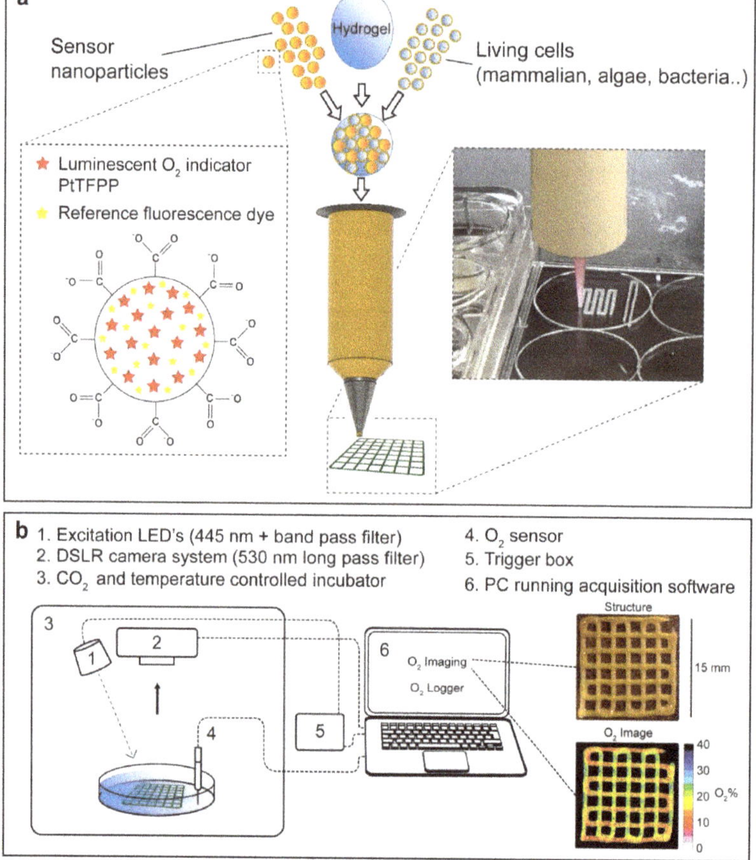

Figure 8. A novel 3D bioprinting approach with hydrogel bioinks functionalized with luminescent nanoparticles. (**a**) Cells and/or nanoparticles, containing the O_2-sensitive luminescent indicator PtTFPP compound and an inert fluorescent coumarin dye, were incorporated into an alginate-based bioink for bioprinting. (**b**) Experimental setup used to image O_2 distribution in bioprinted hydrogel constructs.

2.7. Fluorescence Spectroscopy

Fluorescence spectroscopy utilizes the ability of the targeted molecules to emit light at a different wavelength than the optical excited source (Figure 8) [115]. The information from the emitted photon can then be constructed to produce 2D images with high temporal and spatial resolution [116]. In biology, fluorescence can happen with most biological molecules with the appropriate excitation. However, for specific applications, like cell-based therapy and TE, the emitted signal from the fluorophore molecules can be enhanced through direct or indirect labeling. In the case of indirect labeling, the targeted cells can be engineered, like gene transfection, to express fluorescence proteins like GFP [117]. In indirect labeling, the targeted molecules are attached to certain functional fluorescent molecules which can be activated through an optical excited source [118]. In both cases, the excited wavelength needs to be considered carefully since it can affect the specific photophysical properties of the fluorophores, such as photostability, quantum yield, Stokes shift, and fluorescence lifetime [118].

Nanomaterials have emerged as an effective candidate for direct labeling in fluorescent spectroscopy. Different materials such as quantum dots (QDs) [119], polymers [115,120,121], organic dyes [122], upconversion nanoparticles (UCNPs) [123], and gold nanoparticles (AuNPs) [115] have been considered depending on the specific applications. For cellular tracking, the functional nanomaterials can be attached ex vivo to the targeted protein on the cell surface or they can be infused inside through the process of diffusion or active transport [118]. When excited optically, these nanomaterials can be used to distinguish different performance and functionality of the matrix-embedded cells. In addition to cellular studies, nanomaterials, such as QDs [116] and UCNPs [123], have also been used to label hydrogel scaffold structures to study their degradation. These materials require different excitation wavelengths such as ultraviolet (UV) or visible light for QDs and dyes [115], while UCNPs are more sensitive to NIR wavelengths [123]. However, UV and visible wavelengths have different shortcomings, such as limited penetration depth and potential disintegration of the biological molecules and scaffolds [118]. In addition, even though QDs have excellent optical properties, their suboptimal biocompatibility and biodegradability represent major challenges that hinder their applications in TE [115]. NIR fluorophores can resolve those disadvantages as well as minimize the autofluorescence from cells and tissues, which allows them to be used in a wide range of in vivo tracking of hydrogel degradation (Figure 9) [115].

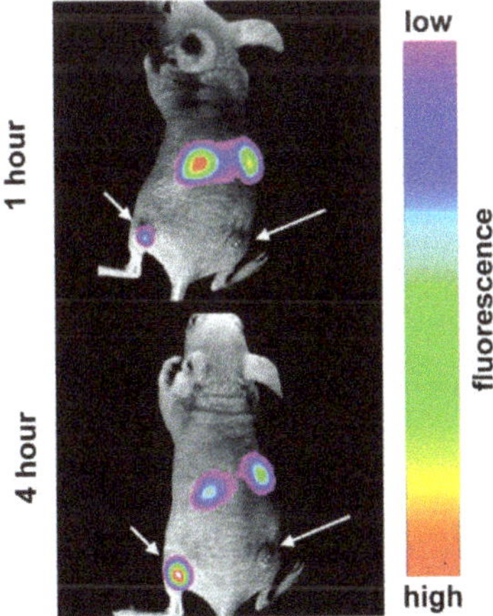

Figure 9. Fluorescence contrast of tumor growth when the mouse is injected with upconversion nanoparticles. Reproduced with permission from Ref. [115].

In addition to fluorescent properties, certain metallic nanoparticles like gold nanoparticles can have multi-functional properties, contributing to tissue mechanical properties, electrical conductivity and (cell) differentiation, and photothermal effect, when they are integrated into the biopolymer scaffold. Such particles, therefore, are good candidates for multifunctional applications such as cancer detection assays and optically-controlled on–off microfluidic devices [124].

2.8. Optical Coherence Tomography

In addition to fluorescence and bioluminescence, which only provide 2D image information, recent advances in optical imaging technologies have allowed for the 3D visualization of tissue structure through the measurement of the interference and coherence between signals reflected from the object and reference signals, known as optical coherence tomography (OCT) [11]. Due to this unique property, OCT can provide anatomical information of the object with sub-millimeter penetration depth [10]. OCT can be used with a variety of light sources, ranging from NIR to visible light [125,126]. In the field of TE, OCT has been used to investigate the geometrical parameters of 3D scaffold architecture, including porosity, surface area, pore sizes, and pore interconnectivity [127,128], as well as remodeling and degradation of polymer structures for specific applications such as vascular grafts [129]. OCT can also be used to asses cell viability, proliferation, distribution, morphology, and function within a cell-laden hydrogel and scaffold (Figure 9) [130]. Advancement in phase-based OCT has also been able to provide contrast between cells and the surrounding hydrogels, thus allowing to achieve a greater understanding of the cell–ECM interactions [128]. Furthermore, a combination of OCT with Doppler velocimetry has been used to characterize flow in engineered tissues, such as artificial blood vessels, by increasing the obtained contrast compared with conventional OCT [11]. Thereupon, OCT has been widely used in conjunction with automated 3D fabrication techniques, such as 3D bioprinting [131], as a high-resolution, noninvasive, label-free method, enabling cellular imaging at different levels for TE applications (Figure 10).

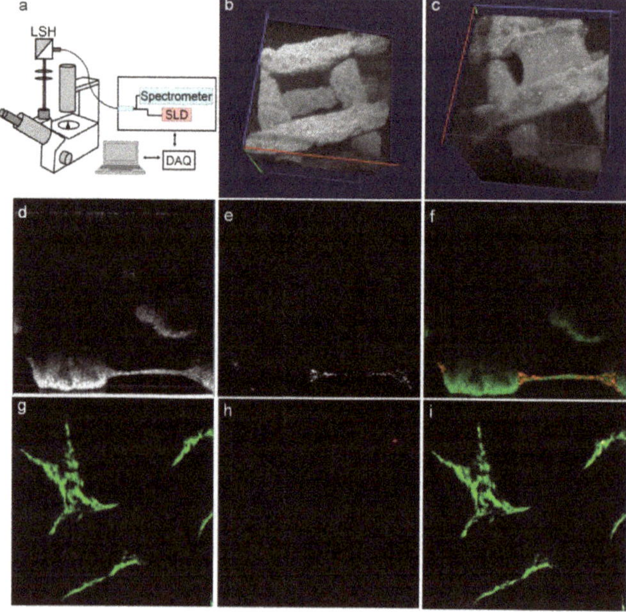

Figure 10. (**a–i**) OCT design and the tracking of cells in a 3D bioprinted scaffold seeded with cells. Reproduced with permission from Ref. [127].

2.9. Photoacoustic Imaging (PAI)

Photoacoustic imaging (PAI) leverages the photoacoustic effect produced by pulsed non-ionizing lasers in tissue to reconstruct an image. For this method, the pulsed laser energy causes heat-induced, elastic tissue expansion, of which emits ultrasonic waves in the MHz range. Via ultrasound transducers, these waves are detected and electronically processed to output a final picture (Figure 11) [132]. PAI diverges based on the acquisition method into photoacoustic microscopy (PAM—focused scanning) and photoacoustic tomography (PACT—inverse reconstruction) [133]. Overall biomedical applications of this technique vary from visualizing macroscopic structures (e.g., small animals to tissues) to microscopic structures (e.g., cells to organelles), with associated contrast agents [133,134]. The advantages of PAI include its non-invasiveness, non-destructiveness, macro to microscale versatility, and compatibility with established imaging modalities. Conversely, PAI challenges include merging optical and acoustic signals, limited scanning speed for wide fields of motion and optimized mathematical models across measurement scales [132,133].

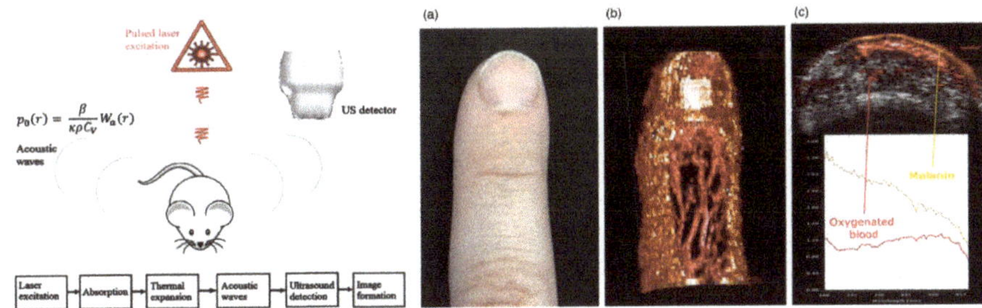

Figure 11. Overview of the physics and processing involved in photoacoustic imaging (PAI) (**left**), and a sample of 2D and 3D vasculatures acquired via PAI from hemoglobin and melanin emissions without contrast agents (**right, a–c**). Reproduced with permission from Ref. [132].

Applications of PAI in TE and 3D (bio)printing are beginning to expand. For instance, Cai et al., compared the resolution of microcomputed tomography to PAM technique in scaffolds of poly(lactic-co-glycolic acid) incorporated with single-walled carbon nanotubes. They demonstrated commensurate porosity measurements under physiological conditions [134]. In another study, acoustic and physiomechanical properties of 3D bioprinted poly-(ethylene glycol)-diacrylate scaffolds were quantified using an ultrasound pulse echo technique [135]. Hu and Wang have shown micrometer-level resolutions of microvasculatures capable of both capturing geometric and hemodynamic information, such as blood oxygenation [136]. This can be extremely useful in monitoring the oxygenation of implanted TE constructs. Much of PAI research is currently focused on optimizing contrast agents for high fidelity imaging for eventual in vivo applications [137–139].

Contrast agents are not always needed due to the PA emissions of already present hemoglobin and melanin. However, to penetrate beyond 1 mm depth, contrast agents that absorb NIR waves are optimal [140]. Metallic (e.g., gold and copper selenide-gold [141]), organic (e.g., carbon tubes and graphene oxide [138]), and semiconductor (e.g., semiconductor polymers and quantum dots [137]) nanoparticles of varying orientations have been employed. Of the three, organic particles exhibit size-independent properties and improved biocompatibility and biodistribution, especially with surface neutralization via encapsulation. This enables multiplexed imaging with customized organic nanoparticles [140]. Overall, PAI development holds much promise in monitoring the TE systems.

2.10. Multimodal Imaging

Each imaging modality, described above, is associated with certain drawbacks and hence, a single imaging modality may not be utilized to acquire all desired information from the tissue/construct of interest. Multimodal imaging can be used to overcome these limitations by utilizing a combination of imaging modalities to provide better resolution in terms of spatial information, along with functional and molecular information [142]. Current platforms of multimodal imaging strategies explore combinations of CT/positron emission tomography (PET), CT/MRI, MRI/PET, and ultrasound/PA to characterize and monitor TE constructs (Figure 12) [143].

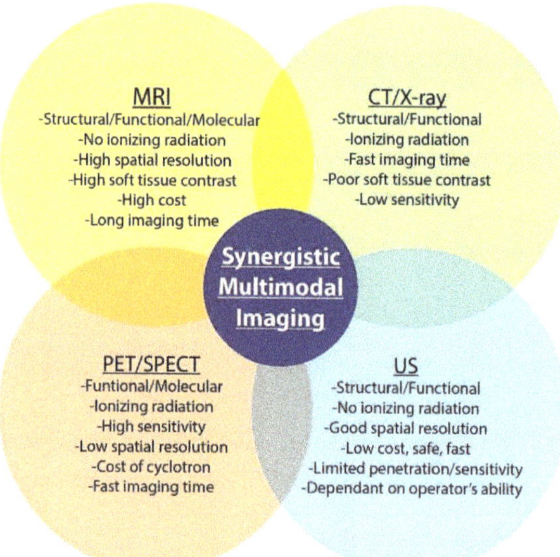

Figure 12. Overview of multimodal imaging applications and their advantages and disadvantages. Reproduced with permission from Ref. [143].

Contrast agents are often used in multimodal imaging, providing reliable detectability. Radioactive isotopes such as 99mTc (t1/2 = 6 h) and 18F (t1/2 = 110 min) are commonly used in single photon emission CT (SPECT) and PET imaging, since both modalities rely on the detection of γ-photons emitted from radioactive isotopes [144,145]. SPECT/CT was utilized to non-invasively monitor bone morphogenetic protein-2 (BMP-2) content and bone formation in composite TE constructs for bone regeneration [146,147]. Kempen et al., created a drug delivery model composed of poly(lactic-co-glycolic acid) embedded into a gelatin hydrogel scaffold over 56 days. A reliable sustained release profile of the 125I-radiolabeled BMP-2 was recorded with SPECT over the full implantation period while in vivo micro-CT detected initial bone formation [146]. However, for multimodal applications, a common contrast agent that each modality can detect is the most favorable. Some nanomaterials such as liposomes, carbon nanotubes, gold nanoparticles, and iron oxide nanoparticles are efficient candidates due to their inherent biocompatibility [17,144,148,149].

Multimodal imaging methods are being increasingly used to track 3D bioprinted tissue constructs. For instance, ultrasound-guided PA imaging technique has shown to have great potential in visualizing the structure, distribution, and retention of microvascular endothelial cells within 3D tissue constructs [150]. Overall, multimodal imaging is dependent on suitable contrast agents that allow for capturing the synergetic properties of each imaging system.

3. Conclusions

Imaging techniques have proven to be indispensable to the advancements in the field of TE and regenerative medicine. In vivo imaging and tracking methods provide vital information about different aspects of engineered tissue constructs post implantation. These features include the 3D geometrical microstructure; the interaction between biological molecules and the scaffold; and the cellular behavior, interaction, and viability within the constructs. Acquiring this information is important to provide feedback to improve the design and fabrication of scaffold systems for different clinical applications, as well as to enhance the understanding of certain cellular processes for cell-based therapies. The individual techniques outlined in this study offer different advantages for the in vivo monitoring of molecules and cells. Different methods have distinct capabilities in tracking different properties of 3D scaffolds. These techniques also suffer from distinct disadvantages which can limit their application in clinical trials. Thus, it is important for the researchers to choose the appropriate imaging modality for specific in vivo studies. The emergence of multimodal imaging has provided an alternative to overcome the shortcomings of the individual imaging techniques, thus enabling a more comprehensive visualization at different levels. Furthermore, progresses in developing various contrast agents for different imaging modalities have enhanced the imaging resolution as well as the ability to combine multifunctional contrast agents as both diagnostics and therapeutics. Finally, advanced imaging techniques can also be combined with new fabrication techniques, such as 3D bioprinting, thus allowing for patient-specific therapeutic applications.

Author Contributions: Conceptualization, V.S. and C.J.G.; Writing-Original Draft Preparation, C.J.G., M.L.T., A.S.T., and A.C.; Writing-Review & Editing, C.J.G., M.M., and V.S.; Project Administration, V.S.

Funding: This research was funded by the National Institute of Health (NIH) grant number R00HL127295 and Emory University School of Medicine (Pediatric Research Alliance Pilot Grant and the Dean's Imagine, Innovate and Impact (I3) Research Award). Carmen J. Gil is supported by the National Science Foundation (NSF) Graduate Research Fellowship under Grant No. DGE-1650044.

Conflicts of Interest: The authors declare no conflict of interest.

References

1. Zhang, Y.S.; Yao, J. Imaging Biomaterial-Tissue Interactions. *Trends Biotechnol.* **2015**, *40*, 1291–1296. [CrossRef] [PubMed]
2. Khan, F.; Tanaka, M. Designing Smart Biomaterials for Tissue Engineering. *Int. J. Mol. Sci.* **2018**, *19*, 17. [CrossRef] [PubMed]
3. Lee, E.J.; Kasper, F.K.; Mikos, A.G. Biomaterials for Tissue Engineering. *Ann. Biomed. Eng.* **2014**, *42*, 323–337. [CrossRef] [PubMed]
4. Zhang, X.; Zhang, Y. Tissue Engineering Applications of Three-Dimensional Bioprinting. *Cell Biochem. Biophys.* **2015**, *72*, 777–782. [CrossRef] [PubMed]
5. Hunsberger, J.; Harrysson, O.; Shirwaiker, R.; Starly, B.; Wysk, R.; Cohen, P.; Allickson, J.; Yoo, J.; Atala, A. Manufacturing Road Map for Tissue Engineering and Regenerative Medicine Technologies. *Stem Cells Transl. Med.* **2015**, *4*, 130–135. [CrossRef] [PubMed]
6. Hu, J.B.; Tomov, M.L.; Buikema, J.W.; Chen, C.; Mahmoudi, M.; Wu, S.M.; Serpooshan, V. Cardiovascular tissue bioprinting: Physical and chemical processes. *Appl. Phys. Rev.* **2018**, *5*, 041106. [CrossRef]
7. Serpooshan, V.; Mahmoudi, M.; Hu, D.A.; Hu, J.B.; Wu, S.M. Bioengineering cardiac constructs using 3D printing. *J. 3D Print. Med.* **2017**, *1*, 123–139. [CrossRef]
8. Serpooshan, V.; Hu, J.B.; Chirikian, O.; Hu, D.A.; Mahmoudi, M.; Wu, S.M. Chapter 8—4D Printing of Actuating Cardiac Tissue. In *3D Printing Applications in Cardiovascular Medicine*; Al'Aref, S.J., Mosadegh, B., Dunham, S., Min, J.K., Eds.; Academic Press: Boston, MA, USA, 2018; pp. 153–162.
9. Appel, A.A.; Anastasio, M.A.; Larson, J.C.; Brey, E.M. Imaging challenges in biomaterials and tissue engineering. *Biomaterials* **2013**, *34*, 6615–6630. [CrossRef] [PubMed]
10. Nam, S.Y.; Ricles, L.M.; Suggs, L.J.; Emelianov, S.Y. Imaging strategies for tissue engineering applications. *Tissue Eng. Part B Rev.* **2014**, *21*, 88–102. [CrossRef]
11. Teodori, L.; Crupi, A.; Costa, A.; Diaspro, A.; Melzer, S.; Tarnok, A. Three-dimensional imaging technologies: A priority for the advancement of tissue engineering and a challenge for the imaging community. *J. Biophotonics* **2017**, *10*, 24–45. [CrossRef]
12. Stacy, M.R.; Sinusas, A.J. Emerging Imaging Modalities in Regenerative Medicine. *Curr. Pathobiol. Rep.* **2015**, *3*, 27–36. [CrossRef] [PubMed]

13. Willadsen, M.; Chaise, M.; Yarovoy, I.; Zhang, A.Q.; Parashurama, N. Engineering molecular imaging strategies for regenerative medicine. *Bioeng. Transl. Med.* **2018**, *3*, 232–255. [CrossRef] [PubMed]
14. Maniam, S.; Szklaruk, J. Magnetic resonance imaging: Review of imaging techniques and overview of liver imaging. *World J. Radiol.* **2010**, *2*, 309–322. [CrossRef] [PubMed]
15. Hartwig, V.; Giovannetti, G.; Vanello, N.; Lombardi, M.; Landini, L.; Simi, S. Biological effects and safety in magnetic resonance imaging: A review. *Int. J. Environ. Res. Public Health* **2009**, *6*, 1778–1798. [CrossRef] [PubMed]
16. Fu, F.; Qin, Z.; Xu, C.; Chen, X.Y.; Li, R.X.; Wang, L.N.; Peng, D.W.; Sun, H.T.; Tu, Y.; Chen, C.; et al. Magnetic resonance imaging-three-dimensional printing technology fabricates customized scaffolds for brain tissue engineering. *Neural Regener. Res.* **2017**, *12*, 614–622.
17. Mahmoudi, M.; Zhao, M.; Matsuura, Y.; Laurent, S.; Yang, P.C.; Bernstein, D.; Ruiz-Lozano, P.; Serpooshan, V. Infection-resistant MRI-visible scaffolds for tissue engineering applications. *Bioimpacts* **2016**, *6*, 111–115. [CrossRef] [PubMed]
18. Wei, K.; Serpooshan, V.; Hurtado, C.; Diez-Cunado, M.; Zhao, M.; Maruyama, S.; Zhu, W.; Fajardo, G.; Noseda, M.; Nakamura, K.; et al. Epicardial FSTL1 reconstitution regenerates the adult mammalian heart. *Nature* **2015**, *525*, 479–485. [CrossRef] [PubMed]
19. Melchels, F.P.W.; Domingos, M.A.N.; Klein, T.J.; Malda, J.; Bartolo, P.J.; Hutmacher, D.W. Additive manufacturing of tissues and organs. *Prog. Polym. Sci.* **2012**, *37*, 1079–1104. [CrossRef]
20. Tegafaw, T.; Xu, W.; Ahmad, M.W.; Baeck, J.S.; Chang, Y.; Bae, J.E.; Chae, K.S.; Kim, T.J.; Lee, G.H. Dual-mode T1 and T2 magnetic resonance imaging contrast agent based on ultrasmall mixed gadolinium-dysprosium oxide nanoparticles: Synthesis, characterization, and in vivo application. *Nanotechnology* **2016**, *26*, 365102. [CrossRef]
21. Krourdan, O.; Ribot, E.; Fricain, J.C.; Devillard, R.; Miraux, S. Magnetic Resonance Imaging for tracking cellular patterns obtained by Laser-Assisted Bioprinting. *Sci. Rep.* **2018**, *8*, 15777. [CrossRef]
22. Choi, J.; Kim, K.; Kim, T.; Liu, G.; Bar-Shir, A.; Hyeon, T.; McMahon, M.T.; Bulte, J.W.; Fisher, J.P.; Gilad, A.A. Multimodal imaging of sustained drug release from 3-D poly(propylene fumarate) (PPF) scaffolds. *J. Control. Release* **2011**, *156*, 239–245. [CrossRef]
23. Kang, X.; Yang, D.; Dai, Y.; Shang, M.; Cheng, Z.; Zhang, X.; Lian, H.; Ma, P.; Lin, J. Poly(acrylic acid) modified lanthanide-doped GdVO4 hollow spheres for up-conversion cell imaging, MRI and pH-dependent drug release. *Nanoscale* **2013**, *5*, 253–261. [CrossRef] [PubMed]
24. Berdichevski, A.; Yameen, H.S.; Dafni, H.; Neeman, M.; Seliktar, D. Using bimodal MRI/fluorescence imaging to identify host angiogenic response to implants. *Proc. Natl. Acad. Sci. USA* **2015**, *112*, 5147–5152. [CrossRef] [PubMed]
25. Zadpoor, A.A.; Malda, J. Additive Manufacturing of Biomaterials, Tissues, and Organs. *Ann. Biomed. Eng.* **2017**, *45*, 1–11. [CrossRef] [PubMed]
26. Liu, L.; Ye, Q.; Wu, Y.; Hsieh, W.Y.; Chen, C.L.; Shen, H.H.; Wang, S.J.; Zhang, H.; Hitchens, T.K.; Ho, C. Tracking T-cells in vivo with a new nano-sized MRI contrast agent. *Nanomedicine* **2012**, *8*, 1345–1354. [CrossRef] [PubMed]
27. Mahmoudi, M.; Yu, M.; Serpooshan, V.; Wu, J.C.; Langer, R.; Lee, R.T.; Karp, J.M.; Farokhzad, O.C. Multiscale technologies for treatment of ischemic cardiomyopathy. *Nat. Nanotechnol.* **2017**, *12*, 845–855. [CrossRef] [PubMed]
28. Serpooshan, V.; Zhao, M.; Metzler, S.A.; Wei, K.; Shah, P.B.; Wang, A.; Mahmoudi, M.; Malkovskiy, A.V.; Rajadas, J.; Butte, M.J.; et al. The effect of bioengineered acellular collagen patch on cardiac remodeling and ventricular function post myocardial infarction. *Biomaterials* **2013**, *34*, 9048–9055. [CrossRef]
29. Chen, Z.; Yan, C.; Yan, S.; Liu, Q.; Hou, M.; Xu, Y.; Guo, R. Non-invasive monitoring of in vivo hydrogel degradation and cartilage regeneration by multiparametric MR imaging. *Theranostics* **2018**, *8*, 1146–1158. [CrossRef] [PubMed]
30. Proulx, M.; Aubin, K.; Lagueux, J.; Audet, P.; Auger, M.; Fortin, M.-A.; Fradette, J. Magnetic Resonance Imaging of Human Tissue-Engineered Adipose Substitutes. *Tissue Eng. Part C Methods* **2014**, *21*, 693–704. [CrossRef] [PubMed]

31. Constantinides, C.; Basnett, P.; Lukasiewicz, B.; Carnicer, R.; Swider, E.; Majid, Q.A.; Srinivas, M.; Carr, C.A.; Roy, I. In Vivo Tracking and 1H/19F Magnetic Resonance Imaging of Biodegradable Polyhydroxyalkanoate/Polycaprolactone Blend Scaffolds Seeded with Labeled Cardiac Stem Cells. *ACS Appl. Mater. Interfaces* **2018**, *10*, 25056–25068. [CrossRef]
32. Kedziorek, D.A.; Kraitchman, D.L. Superparamagnetic iron oxide labeling of stem cells for MRI tracking and delivery in cardiovascular disease. *Methods Mol. Biol.* **2010**, *660*, 171–183. [PubMed]
33. Bulte, J.W. In vivo MRI cell tracking: Clinical studies. *Am. J. Roentgenol.* **2009**, *193*, 314–325. [CrossRef] [PubMed]
34. Tremblay, M.L.; Davis, C.; Bowen, C.V.; Stanley, O.; Parsons, C.; Weir, G.; Karkada, M.; Stanford, M.M.; Brewer, K.D. Using MRI cell tracking to monitor immune cell recruitment in response to a peptide-based cancer vaccine. *Magn. Reson. Med.* **2018**, *80*, 304–316. [CrossRef] [PubMed]
35. Du, Y.; Lai, P.T.; Leung, C.H.; Pong, P.W.T. Design of superparamagnetic nanoparticles for magnetic particle imaging (MPI). *Int. J. Mol. Sci.* **2013**, *14*, 18682–18710. [CrossRef] [PubMed]
36. Bauer, L.M.; Situ, S.F.; Griswold, M.A.; Samia, A.C.S. Magnetic Particle Imaging Tracers: State-of-the-Art and Future Directions. *J. Phys. Chem. Lett.* **2015**, *6*, 2509–2517. [CrossRef]
37. Panagiotopoulos, N.; Vogt, F.; Barkhausen, J.; Buzug, T.M.; Duschka, R.L.; Ldtke-Buzug, K.; Ahlborg, M.; Bringout, G.; Debbeler, C.; Grser, M.; et al. Magnetic particle imaging: Current developments and future directions. *Int. J. Nanomed.* **2015**, *10*, 3097. [CrossRef]
38. Bietenbeck, M.; Florian, A.; Faber, C.; Sechtem, U.; Yilmaz, A. Remote magnetic targeting of iron oxide nanoparticles for cardiovascular diagnosis and therapeutic drug delivery: Where are we now? *Int. J. Nanomed.* **2016**, *11*, 3191–3203.
39. Haegele, J.; Vaalma, S.; Panagiotopoulos, N.; Barkhausen, J.; Vogt, F.M.; Borgert, J.; Rahmer, J. Multi-color magnetic particle imaging for cardiovascular interventions. *Phys. Med. Biol.* **2016**, *61*, N415–N426. [CrossRef]
40. Vogel, P.; Rckert, M.A.; Klauer, P.; Kullmann, W.H.; Jakob, P.M.; Behr, V.C. First in vivo traveling wave magnetic particle imaging of a beating mouse heart. *Phys. Med. Biol.* **2016**, *61*, 6620–6634. [CrossRef]
41. Vaalma, S.; Rahmer, J.; Panagiotopoulos, N.; Duschka, R.L.; Borgert, J.; Barkhausen, J.; Vogt, F.M.; Haegele, J. Magnetic Particle Imaging (MPI): Experimental quantification of vascular stenosis using stationary stenosis phantoms. *PLoS ONE* **2017**, *12*, e0168902.
42. Zhou, X.Y.; Tay, Z.W.; Chandrasekharan, P.; Yu, E.Y.; Hensley, D.W.; Orendorff, R.; Jeffris, K.E.; Mai, D.; Zheng, B.; Goodwill, P.W.; et al. Magnetic particle imaging for radiation-free, sensitive and high-contrast vascular imaging and cell tracking. *Curr. Opin. Chem. Biol.* **2018**, *45*, 131–138. [CrossRef] [PubMed]
43. Hachani, R.; Lowdell, M.; Birchall, M.; Thanh, N.T.K. Tracking stem cells in tissue-engineered organs using magnetic nanoparticles. *Nanoscale* **2013**, *5*, 11362–11373. [CrossRef]
44. Zheng, B.; Vazin, T.; Goodwill, P.W.; Conway, A.; Verma, A.; Saritas, E.U.; Schaffer, D.; Conolly, S.M. Magnetic particle imaging tracks the long-term fate of in vivo neural cell implants with high image contrast. *Sci. Rep.* **2015**, *5*, 14055. [CrossRef] [PubMed]
45. Zheng, B.; von See, M.P.; Yu, E.; Gunel, B.; Lu, K.; Vazin, T.; Schaffer, D.V.; Goodwill, P.W.; Conolly, S.M. Quantitative magnetic particle imaging monitors the transplantation, biodistribution, and clearance of stem cells in vivo. *Theranostics* **2016**, *6*, 291–301. [CrossRef] [PubMed]
46. Jasmin; de Souza, G.T.; Louzada, R.A.; Rosado-de-Castro, P.H.; Mendez-Otero, R.; de Carvalho, A.C.C. Tracking stem cells with superparamagnetic iron oxide nanoparticles: Perspectives and considerations. *Int. J. Nanomed.* **2017**, *12*, 779–793. [CrossRef] [PubMed]
47. Gu, L.; Li, X.; Jiang, J.; Guo, G.; Wu, H.; Wu, M.; Zhu, H. Stem cell tracking using effective self-assembled peptide-modified superparamagnetic nanoparticles. *Nanoscale* **2018**, *10*, 15967–15979. [CrossRef] [PubMed]
48. Tay, Z.W.; Chandrasekharan, P.; Zhou, X.Y.; Yu, E.; Zheng, B.; Conolly, S. In vivo tracking and quantification of inhaled aerosol using magnetic particle imaging towards inhaled therapeutic monitoring. *Theranostics* **2018**, *8*, 3676–3687. [CrossRef]
49. Schwenk, M.H. Ferumoxytol: A new intravenous iron preparation for the treatment of iron deficiency anemia in patients with chronic kidney disease. *Pharmacotherapy* **2010**, *30*, 70–79. [CrossRef]
50. Reimer, P.; Balzer, T. Ferucarbotran (Resovist): A new clinically approved RES-specific contrast agent for contrast-enhanced MRI of the liver: Properties, clinical development, and applications. *Eur. Radiol.* **2003**, *13*, 1266–1276.

51. Wang, Y.X. Current status of superparamagnetic iron oxide contrast agents for liver magnetic resonance imaging. *World J. Gastroenterol.* **2015**, *21*, 13400–13402. [CrossRef]
52. Teshome, M.; Wei, C.; Hunt, K.K.; Thompson, A.; Rodriguez, K.; Mittendorf, E.A. Use of a Magnetic Tracer for Sentinel Lymph Node Detection in Early-Stage Breast Cancer Patients: A Meta-analysis. *Ann. Surg. Oncol.* **2016**, *23*, 1508–1514. [CrossRef] [PubMed]
53. Chang, L.; Liu, X.L.; di Fan, D.; Miao, Y.Q.; Zhang, H.; Ma, H.P.; Liu, Q.Y.; Ma, P.; Xue, W.M.; Luo, Y.E.; et al. The efficiency of magnetic hyperthermia and in vivo histocompatibility for human-like collagen protein-coated magnetic nanoparticles. *Int. J. Nanomed.* **2016**, *11*, 1175–1185.
54. Bulte, J.W.M.; Walczak, P.; Janowski, M.; Krishnan, K.M.; Arami, H.; Halkola, A.; Gleich, B.; Rahmer, J. Quantitative "Hot-Spot" Imaging of Transplanted Stem Cells Using Superparamagnetic Tracers and Magnetic Particle Imaging. *Tomography* **2015**, *1*, 91–97. [PubMed]
55. Thakor, A.S.; Jokerst, J.V.; Ghanouni, P.; Campbell, J.L.; Mittra, E.; Gambhir, S.S. Clinically Approved Nanoparticle Imaging Agents. *J. Nucl. Med.* **2016**, *57*, 1833–1837. [CrossRef] [PubMed]
56. Laperle, C.M.; Hamilton, T.J.; Wintermeyer, P.; Walker, E.J.; Shi, D.; Anastasio, M.A.; Derdak, Z.; Wands, J.R.; Diebold, G.; Rose-Petruck, C. Low density contrast agents for x-ray phase contrast imaging: The use of ambient air for x-ray angiography of excised murine liver tissue. *Phys. Med. Biol.* **2008**, *53*, 6911–6923. [CrossRef]
57. Murata, H.; Oka, Y.; Aoki, K.; Maeda, S.; Yamashita, K.; Terai, H.; Miyamoto, T. Digital subtraction angiography of cardiovascular abnormalities using FCR, a new X-ray diagnostic system. *Kyobu Geka* **1985**, *38*, 530–534.
58. Rhee, T.K.; Park, J.K.; Cashen, T.A.; Shin, W.; Schirf, B.E.; Gehl, J.A.; Larson, A.C.; Carr, J.C.; Li, D.; Carroll, T.J.; et al. Comparison of intraarterial MR angiography at 3.0 T with X-ray digital subtraction angiography for detection of renal artery stenosis in swine. *J. Vasc. Interv. Radiol.* **2006**, *17*, 1131–1137. [CrossRef]
59. Mu, C.L.; Luo, L.H.; Wang, B.P. Comparative study of color Doppler flow imaging with X-ray angiography in the diagnosis of subcutaneous soft-tissue hemangioma. *Nan Fang Yi Ke Da Xue Xue Bao* **2010**, *30*, 2770–2771.
60. Johansson, L.O.; Nolan, M.M.; Taniuchi, M.; Fischer, S.E.; Wickline, S.A.; Lorenz, C.H. High-resolution magnetic resonance coronary angiography of the entire heart using a new blood-pool agent, NC100150 injection: Comparison with invasive x-ray angiography in pigs. *J. Cardiovasc. Magn. Reson.* **1999**, *1*, 139–143. [CrossRef]
61. Geva, T.; Greil, G.F.; Marshall, A.C.; Landzberg, M.; Powell, A.J. Gadolinium-enhanced 3-dimensional magnetic resonance angiography of pulmonary blood supply in patients with complex pulmonary stenosis or atresia: Comparison with x-ray angiography. *Circulation* **2002**, *106*, 473–478. [CrossRef]
62. Anderson, C.M.; Saloner, D.; Lee, R.E.; Griswold, V.J.; Shapeero, L.G.; Rapp, J.H.; Nagarkar, S.; Pan, X.; Gooding, G.A. Assessment of carotid artery stenosis by MR angiography: Comparison with x-ray angiography and color-coded Doppler ultrasound. *Am. J. Neuroradiol.* **1992**, *13*, 989–1003. [PubMed]
63. Wutke, R.; Lang, W.; Fellner, C.; Janka, R.; Denzel, C.; Lell, M.; Bautz, W.; Fellner, F.A. High-resolution, contrast-enhanced magnetic resonance angiography with elliptical centric k-space ordering of supra-aortic arteries compared with selective X-ray angiography. *Stroke* **2002**, *33*, 1522–1529. [CrossRef] [PubMed]
64. Hartung, M.P.; Grist, T.M.; Francois, C.J. Magnetic resonance angiography: Current status and future directions. *J. Cardiovasc. Magn. Reson.* **2011**, *13*, 19. [CrossRef] [PubMed]
65. Gupta, P.N.; Basheer, A.S.; Sukumaran, G.G.; Padmajan, S.; Praveen, S.; Velappan, P.; Nair, B.U.; Nair, S.G.; Kunjuraman, U.K.; Madthipat, U.; et al. Femoral artery pseudoaneurysm as a complication of angioplasty. How can it be prevented? *Heart Asia* **2013**, *5*, 144–147. [CrossRef] [PubMed]
66. Cox, S.C.; Thornby, J.A.; Gibbons, G.J.; Williams, M.A.; Mallick, K.K. 3D printing of porous hydroxyapatite scaffolds intended for use in bone tissue engineering applications. *Mater. Sci. Eng. C* **2015**, *47*, 237–247. [CrossRef] [PubMed]
67. Martin, J.T.; Milby, A.H.; Ikuta, K.; Poudel, S.; Pfeifer, C.G.; Elliott, D.M.; Smith, H.E.; Mauck, R.L. A radiopaque electrospun scaffold for engineering fibrous musculoskeletal tissues: Scaffold characterization and in vivo applications. *Acta Biomater.* **2015**, *26*, 97–104. [CrossRef] [PubMed]
68. Osorio, D.A.; Lee, B.E.J.; Kwiecien, J.M.; Wang, X.; Shahid, I.; Hurley, A.L.; Cranston, E.D.; Grandfield, K. Cross-linked cellulose nanocrystal aerogels as viable bone tissue scaffolds. *Acta Biomater.* **2019**, *87*, 152–165. [CrossRef] [PubMed]

69. Ribeiro, V.P.; Pina, S.; Costa, J.B.; Cengiz, I.F.; Garca-Fernndez, L.; Fernndez-Gutirrez, M.D.M.; Paiva, O.C.; Oliveira, A.L.; San-Romn, J.; Oliveira, J.M.; et al. Enzymatically Cross-Linked Silk Fibroin-Based Hierarchical Scaffolds for Osteochondral Regeneration. *ACS Appl. Mater. Interfaces* **2019**, *11*, 3781–3799. [CrossRef]
70. Shepherd, D.V.; Shepherd, J.H.; Best, S.M.; Cameron, R.E. 3D imaging of cells in scaffolds: Direct labelling for micro CT. *J. Mater. Sci. Mater. Med.* **2018**, *29*, 86. [CrossRef]
71. Zidek, J.; Vojtova, L.; Abdel-Mohsen, A.M.; Chmelik, J.; Zikmund, T.; Brtnikova, J.; Jakubicek, R.; Zubal, L.; Jan, J.; Kaiser, J. Accurate micro-computed tomography imaging of pore spaces in collagen-based scaffold. *J. Mater. Sci. Mater. Med.* **2016**, *27*, 110. [CrossRef]
72. Gmez, S.; Vlad, M.D.; Lpez, J.; Fernndez, E. Design and properties of 3D scaffolds for bone tissue engineering. *Acta Biomater.* **2016**, *42*, 341–350. [CrossRef] [PubMed]
73. Izadifar, Z.; Honaramooz, A.; Wiebe, S.; Belev, G.; Chen, X.; Chapman, D. Low-dose phase-based X-ray imaging techniques for in situ soft tissue engineering assessments. *Biomaterials* **2016**, *82*, 151–167. [CrossRef] [PubMed]
74. Sonnaert, M.; Kerckhofs, G.; Papantoniou, I.; Van Vlierberghe, S.; Boterberg, V.; Dubruel, P.; Luyten, F.P.; Schrooten, J.; Geris, L. Multifactorial optimization of contrast-enhanced nanofocus computed tomography for quantitative analysis of neo-tissue formation in tissue engineering constructs. *PLoS ONE* **2015**, *10*, e0130227. [CrossRef] [PubMed]
75. Appel, A.A.; Larson, J.C.; Garson, A.B.; Guan, H.; Zhong, Z.; Nguyen, B.N.B.; Fisher, J.P.; Anastasio, M.A.; Brey, E.M. X-ray phase contrast imaging of calcified tissue and biomaterial structure in bioreactor engineered tissues. *Biotechnol. Bioeng.* **2015**, *112*, 612–620. [CrossRef] [PubMed]
76. Appel, A.A.; Larson, J.C.; Jiang, B.; Zhong, Z.; Anastasio, M.A.; Brey, E.M. X-ray Phase Contrast Allows Three Dimensional, Quantitative Imaging of Hydrogel Implants. *Ann. Biomed. Eng.* **2016**, *44*, 773–781. [CrossRef] [PubMed]
77. Meir, R.; Shamalov, K.; Betzer, O.; Motiei, M.; Horovitz-Fried, M.; Yehuda, R.; Popovtzer, A.; Popovtzer, R.; Cohen, C.J. Nanomedicine for Cancer Immunotherapy: Tracking Cancer-Specific T-Cells in Vivo with Gold Nanoparticles and CT Imaging. *ACS Nano* **2015**, *9*, 6363–6372. [CrossRef] [PubMed]
78. Bernstein, A.L.; Dhanantwari, A.; Jurcova, M.; Cheheltani, R.; Naha, P.C.; Ivanc, T.; Shefer, E.; Cormode, D.P. Improved sensitivity of computed tomography towards iodine and gold nanoparticle contrast agents via iterative reconstruction methods. *Sci. Rep.* **2016**, *6*, 26177. [CrossRef]
79. Khademi, S.; Sarkar, S.; Shakeri-Zadeh, A.; Attaran, N.; Kharrazi, S.; Ay, M.R.; Ghadiri, H. Folic acid-cysteamine modified gold nanoparticle as a nanoprobe for targeted computed tomography imaging of cancer cells. *Mater. Sci. Eng. C* **2018**, *89*, 182–193. [CrossRef]
80. Celikkin, N.; Mastrogiacomo, S.; Walboomers, X.; Swieszkowski, W. Enhancing X-ray Attenuation of 3D Printed Gelatin Methacrylate (GelMA) Hydrogels Utilizing Gold Nanoparticles for Bone Tissue Engineering Applications. *Polymers* **2019**, *11*, 367. [CrossRef]
81. Finamore, T.A.; Curtis, T.E.; Tedesco, J.V.; Grandfield, K.; Roeder, R.K. Nondestructive, longitudinal measurement of collagen scaffold degradation using computed tomography and gold nanoparticles. *Nanoscale* **2019**, *11*, 4345–4354. [CrossRef]
82. Wang, Y.; Xiong, Z.; He, Y.; Zhou, B.; Qu, J.; Shen, M.; Shi, X.; Xia, J. Optimization of the composition and dosage of PEGylated polyethylenimine-entrapped gold nanoparticles for blood pool, tumor, and lymph node CT imaging. *Mater. Sci. Eng. C* **2018**, *83*, 9–16. [CrossRef] [PubMed]
83. Cheheltani, R.; Ezzibdeh, R.M.; Chhour, P.; Pulaparthi, K.; Kim, J.; Jurcova, M.; Hsu, J.C.; Blundell, C.; Litt, H.I.; Ferrari, V.A.; et al. Tunable, biodegradable gold nanoparticles as contrast agents for computed tomography and photoacoustic imaging. *Biomaterials* **2016**, *102*, 87–97. [CrossRef] [PubMed]
84. Kim, J.; Chhour, P.; Hsu, J.; Litt, H.I.; Ferrari, V.A.; Popovtzer, R.; Cormode, D.P. Use of Nanoparticle Contrast Agents for Cell Tracking with Computed Tomography. *Bioconjugate Chem.* **2017**, *28*, 1581–1597. [CrossRef] [PubMed]
85. Lakin, B.A.; Patel, H.; Holland, C.; Freedman, J.D.; Shelofsky, J.S.; Snyder, B.D.; Stok, K.S.; Grinstaff, M.W. Contrast-enhanced CT using a cationic contrast agent enables non-destructive assessment of the biochemical and biomechanical properties of mouse tibial plateau cartilage. *J. Orthop. Res.* **2016**, *34*, 1130–1138. [CrossRef] [PubMed]

86. Chae, K.S.; Ahmad, M.W.; Kim, T.J.; Lee, G.H.; Xu, W.; Baeck, J.S.; Kim, S.J.; Park, J.A.; Bae, J.E.; Chang, Y. Synthesis of nanoparticle CT contrast agents: In vitro and in vivo studies. *Sci. Technol. Adv. Mater.* **2015**, *16*, 055003.
87. Torchilin, V.P. Polymeric contrast agents for medical imaging. *Curr. Pharm. Biotechnol.* **2000**, *1*, 183–215. [CrossRef] [PubMed]
88. Boll, H.; Nittka, S.; Doyon, F.; Neumaier, M.; Marx, A.; Kramer, M.; Groden, C.; Brockmann, M.A. Micro-CT based experimental liver imaging using a nanoparticulate contrast agent: A longitudinal study in mice. *PLoS One* **2011**, *6*, e25692. [CrossRef] [PubMed]
89. Emmert, M.Y.; Schmitt, B.A.; Loerakker, S.; Sanders, B.; Spriestersbach, H.; Fioretta, E.S.; Bruder, L.; Brakmann, K.; Motta, S.E.; Lintas, V.; et al. Computational modeling guides tissue-engineered heart valve design for long-term in vivo performance in a translational sheep model. *Sci. Transl. Med.* **2018**, *10*, eaan4587. [CrossRef]
90. Kajbafzadeh, A.M.; Ahmadi Tafti, S.H.; Mokhber-Dezfooli, M.R.; Khorramirouz, R.; Sabetkish, S.; Sabetkish, N.; Rabbani, S.; Tavana, H.; Mohseni, M.J. Aortic valve conduit implantation in the descending thoracic aorta in a sheep model: The outcomes of pre-seeded scaffold. *Int. J. Surg.* **2016**, *28*, 97–105. [CrossRef]
91. Zhu, J.; Wang, G.; Alves, C.S.; Toms, H.; Xiong, Z.; Shen, M.; Rodrigues, J.; Shi, X. Multifunctional Dendrimer-Entrapped Gold Nanoparticles Conjugated with Doxorubicin for pH-Responsive Drug Delivery and Targeted Computed Tomography Imaging. *Langmuir* **2018**, *34*, 12428–12435. [CrossRef]
92. Shikhaliev, P.M. Soft tissue imaging with photon counting spectroscopic CT. *Phys. Med. Biol.* **2015**, *60*, 2453–2474. [CrossRef] [PubMed]
93. Talukdar, Y.; Avti, P.; Sun, J.; Sitharaman, B. Multimodal ultrasound-photoacoustic imaging of tissue engineering scaffolds and blood oxygen saturation in and around the scaffolds. *Tissue Eng. Part C Methods* **2014**, *20*, 440–449. [CrossRef] [PubMed]
94. Deng, C.X.; Hong, X.; Stegemann, J.P. Ultrasound Imaging Techniques for Spatiotemporal Characterization of Composition, Microstructure, and Mechanical Properties in Tissue Engineering. *Tissue Eng. Part B Rev.* **2016**, *22*, 311–321. [CrossRef]
95. Kim, K.; Wagner, W.R. Non-invasive and Non-destructive Characterization of Tissue Engineered Constructs Using Ultrasound Imaging Technologies: A Review. *Ann. Biomed. Eng.* **2016**, *44*, 621–635. [CrossRef] [PubMed]
96. Stukel, J.M.; Goss, M.; Zhou, H.; Zhou, W.; Willits, R.K.; Exner, A.A. Development of a High-Throughput Ultrasound Technique for the Analysis of Tissue Engineering Constructs. *Ann. Biomed. Eng.* **2016**, *44*, 793–802. [CrossRef] [PubMed]
97. Shinohara, M.; Sabra, K.; Gennisson, J.L.; Fink, M.; Tanter, M. Real-time visualization of muscle stiffness distribution with ultrasound shear wave imaging during muscle contraction. *Muscle Nerve* **2010**, *42*, 438–441. [CrossRef] [PubMed]
98. Arda, K.; Ciledag, N.; Aktas, E.; Aribas, B.K.; Kose, K. Quantitative assessment of normal soft-tissue elasticity using shear-wave ultrasound elastography. *Am. J. Roentgenol.* **2011**, *197*, 532–536. [CrossRef] [PubMed]
99. Itoh, A.; Ueno, E.; Tohno, E.; Kamma, H.; Takahashi, H.; Shiina, T.; Yamakawa, M.; Matsumura, T. Breast disease: Clinical application of US elastography for diagnosis. *Radiology* **2006**, *239*, 341–350. [CrossRef]
100. Walker, J.M.; Myers, A.M.; Schluchter, M.D.; Goldberg, V.M.; Caplan, A.I.; Berilla, J.A.; Mansour, J.M.; Welter, J.F. Nondestructive evaluation of hydrogel mechanical properties using ultrasound. *Ann. Biomed. Eng.* **2011**, *39*, 2521–2530. [CrossRef] [PubMed]
101. Zhou, Y. The Application of Ultrasound in 3D Bio-Printing. *Molecules* **2016**, *21*, 590. [CrossRef]
102. Garvin, K.A.; Dalecki, D.; Hocking, D.C. Vascularization of three-dimensional collagen hydrogels using ultrasound standing wave fields. *Ultrasound Med. Biol.* **2011**, *37*, 1853–1864. [CrossRef] [PubMed]
103. Angele, P.; Yoo, J.U.; Smith, C.; Mansour, J.; Jepsen, K.J.; Nerlich, M.; Johnstone, B. Cyclic hydrostatic pressure enhances the chondrogenic phenotype of human mesenchymal progenitor cells differentiated in vitro. *J. Orthop. Res.* **2003**, *21*, 451–457. [CrossRef]
104. Kim, J.E.; Kalimuthu, S.; Ahn, B.C. In Vivo Cell Tracking with Bioluminescence Imaging. *Nucl. Med. Mol. Imaging* **2015**, *49*, 3–10. [CrossRef] [PubMed]
105. Yun, S.H.; Kwok, S.J.J. Light in diagnosis, therapy and surgery. *Nat. Biomed. Eng.* **2017**, *1*, 8. [CrossRef] [PubMed]

106. Arranz, A.; Ripoll, J. Advances in optical imaging for pharmacological studies. *Front. Pharmacol.* **2015**, *6*, 189. [CrossRef]
107. Sharkey, J.; Scarfe, L.; Santeramo, I.; Garcia-Finana, M.; Park, B.K.; Poptani, H.; Wilm, B.; Taylor, A.; Murray, P. Imaging technologies for monitoring the safety, efficacy and mechanisms of action of cell-based regenerative medicine therapies in models of kidney disease. *Eur. J. Pharmacol.* **2016**, *790*, 74–82. [CrossRef] [PubMed]
108. Conradi, L.; Schmidt, S.; Neofytou, E.; Deuse, T.; Peters, L.; Eder, A.; Hua, X.; Hansen, A.; Robbins, R.C.; Beygui, R.E.; et al. Immunobiology of Fibrin-Based Engineered Heart Tissue. *Stem Cells Transl. Med.* **2015**, *4*, 625–631. [CrossRef]
109. Crowley, C.; Butler, C.R.; Camilli, C.; Hynds, R.E.; Kolluri, K.K.; Janes, S.M.; De Coppi, P.; Urbani, L. Non-Invasive Longitudinal Bioluminescence Imaging of Human Mesoangioblasts in Bioengineered Esophagi. *Tissue Eng. Part C Methods* **2019**, *25*, 103–113. [CrossRef]
110. Kuchimaru, T.; Iwano, S.; Kiyama, M.; Mitsumata, S.; Kadonosono, T.; Niwa, H.; Maki, S.; Kizaka-Kondoh, S. A luciferin analogue generating near-infrared bioluminescence achieves highly sensitive deep-tissue imaging. *Nat. Commun.* **2016**, *7*, 11856. [CrossRef]
111. Iwano, S.; Sugiyama, M.; Hama, H.; Watakabe, A.; Hasegawa, N.; Kuchimaru, T.; Tanaka, K.Z.; Takahashi, M.; Ishida, Y.; Hata, J.; et al. Single-cell bioluminescence imaging of deep tissue in freely moving animals. *Science* **2018**, *359*, 935–939. [CrossRef]
112. Peeters, M.; Van Rijn, S.; Vergroesen, P.P.A.; Paul, C.P.; Noske, D.P.; Vandertop, W.P.; Wurdinger, T.; Helder, M.N. Bioluminescence-mediated longitudinal monitoring of adipose-derived stem cells in a large mammal ex vivo organ culture. *Sci. Rep.* **2015**, *5*, 13960. [CrossRef] [PubMed]
113. Trampe, E.; Koren, K.; Akkineni, A.R.; Senwitz, C.; Krujatz, F.; Lode, A.; Gelinsky, M.; Kühl, M. Functionalized Bioink with Optical Sensor Nanoparticles for O2 Imaging in 3D-Bioprinted Constructs. *Adv. Funct. Mater.* **2018**, *28*, 1804411. [CrossRef]
114. Keriquel, V.; Oliveira, H.; Remy, M.; Ziane, S.; Delmond, S.; Rousseau, B.; Rey, S.; Catros, S.; Amedee, J.; Guillemot, F.; et al. In situ printing of mesenchymal stromal cells, by laser-assisted bioprinting, for in vivo bone regeneration applications. *Sci. Rep.* **2017**, *7*, 1778. [CrossRef] [PubMed]
115. Chinen, A.B.; Guan, C.M.; Ferrer, J.R.; Barnaby, S.N.; Merkel, T.J.; Mirkin, C.A. Nanoparticle Probes for the Detection of Cancer Biomarkers, Cells, and Tissues by Fluorescence. *Chem. Rev.* **2015**, *115*, 10530–10574. [CrossRef]
116. Tajika, Y.; Murakami, T.; Iijima, K.; Gotoh, H.; Takahashi-Ikezawa, M.; Ueno, H.; Yoshimoto, Y.; Yorifuji, H. A novel imaging method for correlating 2D light microscopic data and 3D volume data based on block-face imaging. *Sci. Rep.* **2017**, *7*, 3645. [CrossRef] [PubMed]
117. Liu, J.; Hilderink, J.; Groothuis, T.A.M.; Otto, C.; van Blitterswijk, C.A.; de Boer, J. Monitoring nutrient transport in tissue-engineered grafts. *J. Tissue Eng. Regener. Med.* **2015**, *9*, 952–960. [CrossRef] [PubMed]
118. Park, G.K.; I, H.; Kim, G.S.; Hwang, N.S.; Choi, H.S. Optical spectroscopic imaging for cell therapy and tissue engineering. *Appl. Spectrosc. Rev.* **2018**, *53*, 360–375. [CrossRef] [PubMed]
119. Bilen, B.; Gokbulut, B.; Kafa, U.; Heves, E.; Inci, M.N.; Unlu, M.B. Scanning Acoustic Microscopy and Time-Resolved Fluorescence Spectroscopy for Characterization of Atherosclerotic Plaques. *Sci. Rep.* **2018**, *8*, 14378. [CrossRef]
120. Sabapathy, V.; Mentam, J.; Jacob, P.M.; Kumar, S. Noninvasive Optical Imaging and In Vivo Cell Tracking of Indocyanine Green Labeled Human Stem Cells Transplanted at Superficial or In-Depth Tissue of SCID Mice. *Stem Cells Int.* **2015**, *2015*, 606415. [CrossRef]
121. Artzi, N.; Oliva, N.; Puron, C.; Shitreet, S.; Artzi, S.; Bon Ramos, A.; Groothuis, A.; Sahagian, G.; Edelman, E.R. In vivo and in vitro tracking of erosion in biodegradable materials using non-invasive fluorescence imaging. *Nat. Mater.* **2011**, *10*, 704–709.
122. Cen, P.; Chen, J.; Hu, C.; Fan, L.; Wang, J.; Li, L. Noninvasive in-vivo tracing and imaging of transplanted stem cells for liver regeneration. *Stem Cell Res. Ther.* **2016**, *7*, 143. [CrossRef] [PubMed]
123. Dong, Y.; Jin, G.; Ji, C.; He, R.; Lin, M.; Zhao, X.; Li, A.; Lu, T.J.; Xu, F. Non-invasive tracking of hydrogel degradation using upconversion nanoparticles. *Acta Biomater.* **2017**, *55*, 410–419. [CrossRef] [PubMed]
124. Thoniyot, P.; Tan, M.J.; Karim, A.A.; Young, D.J.; Loh, X.J. Nanoparticle-Hydrogel Composites: Concept, Design, and Applications of These Promising, Multi-Functional Materials. *Adv. Sci.* **2015**, *2*, 1400010. [CrossRef] [PubMed]

125. Shu, X.; Beckmann, L.; Zhang, H. Visible-light optical coherence tomography: A review. *J. Biomed. Opt.* **2017**, *22*, 121707. [CrossRef] [PubMed]
126. Kim, J.; Brown, W.; Maher, J.R.; Levinson, H.; Wax, A. Functional optical coherence tomography: Principles and progress. *Phys. Med. Biol.* **2015**, *60*, R211–R237. [CrossRef] [PubMed]
127. Holmes, C.; Tabrizian, M.; Bagnaninchi, P.O. Motility imaging via optical coherence phase microscopy enables label-free monitoring of tissue growth and viability in 3D tissue-engineering scaffolds. *J. Tissue Eng. Regener. Med.* **2015**, *9*, 641–645. [CrossRef] [PubMed]
128. Wang, L.; Xu, M.; Zhang, L.; Zhou, Q.; Luo, L. Automated quantitative assessment of three-dimensional bioprinted hydrogel scaffolds using optical coherence tomography. *Biomed. Opt. Express* **2016**, *7*, 894–910. [CrossRef]
129. Chen, W.; Yang, J.; Liao, W.; Zhou, J.; Zheng, J.; Wu, Y.; Li, D.; Lin, Z. In vitro remodeling and structural characterization of degradable polymer scaffold-based tissue-engineered vascular grafts using optical coherence tomography. *Cell Tissue Res.* **2017**, *370*, 417–426. [CrossRef] [PubMed]
130. Wang, L.; Xu, M.E.; Luo, L.; Zhou, Y.; Si, P. Iterative feedback bio-printing-derived cell-laden hydrogel scaffolds with optimal geometrical fidelity and cellular controllability. *Sci. Rep.* **2018**, *8*, 2802. [CrossRef] [PubMed]
131. Guo, T.; Noshin, M.; Baker, H.B.; Taskoy, E.; Meredith, S.J.; Tang, Q.; Ringel, J.P.; Lerman, M.J.; Chen, Y.; Packer, J.D.; et al. 3D printed biofunctionalized scaffolds for microfracture repair of cartilage defects. *Biomaterials* **2018**, *185*, 219–231. [CrossRef] [PubMed]
132. Avigo, C.; Flori, A.; Armanetti, P.; Di Lascio, N.; Kusmic, C.; Jose, J.; Losi, P.; Soldani, G.; Faita, F.; Menichetti, L. Strategies for non-invasive imaging of polymeric biomaterial in vascular tissue engineering and regenerative medicine using ultrasound and photoacoustic techniques. *Polym. Int.* **2016**, *65*, 734–740. [CrossRef]
133. Wang, L.V.; Yao, J. A practical guide to photoacoustic tomography in the life sciences. *Nat. Methods* **2016**, *13*, 627–638. [CrossRef] [PubMed]
134. Cai, X.; Paratala, B.S.; Hu, S.; Sitharaman, B.; Wang, L.V. Multiscale photoacoustic microscopy of single-walled carbon nanotube-incorporated tissue engineering scaffolds. *Tissue Eng. Part C Methods* **2012**, *18*, 310–317. [CrossRef] [PubMed]
135. Aliabouzar, M.; Zhang, G.L.; Sarkar, K. Acoustic and mechanical characterization of 3D-printed scaffolds for tissue engineering applications. *Biomed. Mater.* **2018**, *13*, 055013. [CrossRef] [PubMed]
136. Hu, S.; Wang, L.V. Photoacoustic imaging and characterization of the microvasculature. *J. Biomed. Opt.* **2010**, *15*, 011101. [CrossRef] [PubMed]
137. Lyu, Y.; Fang, Y.; Miao, Q.; Zhen, X.; Ding, D.; Pu, K. Intraparticle Molecular Orbital Engineering of Semiconducting Polymer Nanoparticles as Amplified Theranostics for in Vivo Photoacoustic Imaging and Photothermal Therapy. *ACS Nano* **2016**, *10*, 4472–4481. [CrossRef] [PubMed]
138. Moon, H.; Kim, H.; Kumar, D.; Kim, H.; Sim, C.; Chang, J.H.; Kim, J.M.; Lim, D.K. Amplified photoacoustic performance and enhanced photothermal stability of reduced graphene oxide coated gold nanorods for sensitive photoacoustic imaging. *ACS Nano* **2015**, *9*, 2711–2719. [CrossRef] [PubMed]
139. Antaris, A.L.; Chen, H.; Cheng, K.; Sun, Y.; Hong, G.; Qu, C.; Diao, S.; Deng, Z.; Hu, X.; Zhang, B.; et al. A small-molecule dye for NIR-II imaging. *Nat. Mater.* **2016**, *15*, 235–242. [CrossRef]
140. Jiang, Y.; Pu, K. Advanced Photoacoustic Imaging Applications of Near-Infrared Absorbing Organic Nanoparticles. *Small* **2017**, *13*, 1700710. [CrossRef]
141. Huang, Q.; Zhang, S.; Zhang, H.; Han, Y.; Liu, H.; Ren, F.; Sun, Q.; Li, Z.; Gao, M. Boosting the Radiosensitizing and Photothermal Performance of Cu_{2-x} Se Nanocrystals for Synergetic Radiophotothermal Therapy of Orthotopic Breast Cancer. *ACS Nano* **2019**, *13*, 1342–1353.
142. Van der Horst, G.; Buijs, J.T.; van der Pluijm, G. Chapter 46—Pre-clinical molecular imaging of "the seed and the soil" in bone metastasis. In *Bone Cancer*, 2nd ed.; Heymann, D., Ed.; Academic Press: San Diego, CA, USA, 2015; pp. 557–570.
143. Rieffel, J.; Chitgupi, U.; Lovell, J.F. Recent Advances in Higher-Order, Multimodal, Biomedical Imaging Agents. *Small* **2015**, *11*, 4445–4461. [CrossRef] [PubMed]
144. Chen, D.; Dougherty, C.A.; Yang, D.; Wu, H.; Hong, H. Radioactive Nanomaterials for Multimodality Imaging. *Tomography* **2016**, *2*, 3–16. [PubMed]
145. Ametamey, S.M.; Honer, M.; Schubiger, P.A. Molecular imaging with PET. *Chem. Rev.* **2008**, *108*, 1501–1516. [CrossRef] [PubMed]

146. Kempen, D.H.; Yaszemski, M.J.; Heijink, A.; Hefferan, T.E.; Creemers, L.B.; Britson, J.; Maran, A.; Classic, K.L.; Dhert, W.J.; Lu, L. Non-invasive monitoring of BMP-2 retention and bone formation in composites for bone tissue engineering using SPECT/CT and scintillation probes. *J. Control. Release* **2009**, *134*, 169–176. [CrossRef] [PubMed]
147. Zhou, M.; Peng, X.; Mao, C.; Tian, J.H.; Zhang, S.W.; Xu, F.; Tu, J.J.; Liu, S.; Hu, M.; Yu, G.Y. The Value of SPECT/CT in Monitoring Prefabricated Tissue-Engineered Bone and Orthotopic rhBMP-2 Implants for Mandibular Reconstruction. *PLoS ONE* **2015**, *10*, e0137167. [CrossRef] [PubMed]
148. Abou, D.S.; Thorek, D.L.; Ramos, N.N.; Pinkse, M.W.; Wolterbeek, H.T.; Carlin, S.D.; Beattie, B.J.; Lewis, J.S. (89)Zr-labeled paramagnetic octreotide-liposomes for PET-MR imaging of cancer. *Pharm. Res.* **2013**, *30*, 878–888. [CrossRef] [PubMed]
149. Alex, S.; Tiwari, A. Functionalized Gold Nanoparticles: Synthesis, Properties and Applications—A Review. *J. Nanosci. Nanotechnol.* **2015**, *15*, 1869–1894. [CrossRef] [PubMed]
150. Nagao, R.J.; Ouyang, Y.; Keller, R.; Nam, S.Y.; Malik, G.R.; Emelianov, S.Y.; Suggs, L.J.; Schmidt, C.E. Ultrasound-guided photoacoustic imaging-directed re-endothelialization of acellular vasculature leads to improved vascular performance. *Acta Biomater.* **2016**, *32*, 35–45. [CrossRef]

MDPI
St. Alban-Anlage 66
4052 Basel
Switzerland
Tel. +41 61 683 77 34
Fax +41 61 302 89 18
www.mdpi.com

Micromachines Editorial Office
E-mail: micromachines@mdpi.com
www.mdpi.com/journal/micromachines

www.ingramcontent.com/pod-product-compliance
Lightning Source LLC
LaVergne TN
LVHW070844170726
843515LV00005B/567
9783039361120